Linguistische
Arbeiten 185

Herausgegeben von Hans Altmann, Herbert E. Brekle, Hans Jürgen Heringer, Christian Rohrer, Heinz Vater und Otmar Werner

Essays on Tensing in English

Vol. I: Reference Time, Tense and Adverbs

Edited by
Alfred Schopf

Max Niemeyer Verlag
Tübingen 1987

CIP-Kurztitelaufnahme der Deutschen Bibliothek

Essays on tensing in English / ed. by Alfred Schopf. – Tübingen : Niemeyer
NE: Schopf, Alfred [Hrsg.]
Vol. 1. Reference time, tense and adverbs. – 1987.
(Linguistische Arbeiten ; 185)
NE: GT

ISBN 3-484-30185-6 ISSN 0344-6727

Printed in Germany. Druck: Weihert-Druck, Darmstadt.

INHALT

Foreword

This volume of essays on the tensing system could not have been written without the aid of the Deutsche Forschungsgemeinschaft, who financed project Scho 126/ 5-1 *Textkonstruktion im Englischen I: Das Englische Verzeitungssystem* (1983-85) and the Wissenschaftliche Gesellschaft at the University of Freiburg, who covered travel expenses and enabled contributors to this volume living outside Freiburg to take part in discussions and presentations of our research. We are grateful to these two institutions for their general support. Thanks are also due to those who undertook the task of producing the finished version from our various typescripts and manuscripts, above all to Katharina Nanov-Schwehr and Damian Padberg, who supervised the typing and prepared the bibliography and to the typists, Lydia Zillert, Ute Kongsbak, Diana Pounsford and Frauke Liebschner.

Introduction

This is the first of two volumes of articles which are the product of work embarked upon during a project sponsored for two years by the Deutsche Forschungsgemeinschaft. The articles are the outcome of lengthy and at times heated discussions among members of the project group and to some extent represent various stages in the development of our ideas.

No attempt was made to impose a 'consensus view' upon the various approaches taken here; terminological divergences remain, too. Nevertheless, we hope that the articles can serve as a useful introduction to an area of research which continues to attract considerable attention from scholars.

The starting point for our considerations of problems connected with the English tensing system was a book by the editor of the present volume, *Das Verzeitungssystem des Englischen und seine Textfunktion*, which appeared in 1984. Hans Reichenbach's trichotomous approach to the tensing systems of natural language as presented in *Elements of Symbolic Logic* (1947) remained seminal.

The articles return repeatedly to consideration of Reichenbach's ideas and to examination of proposals developing his ideas made in more recent scholarship.

The first article, Elizabeth Couper-Kuhlen's "Temporal Relations and Reference Times in Narrative Discourse" considers extensions of Reichenbach's tense analysis (primarily sentence-based) to discourse, in particular to narration, and discusses the need for reference time at the level of the text. A critical examination is made of two models which both advocate automatic progression of the reference time in narration, triggered by the event type of the propositions involved in successive sentences (iconic progression). The first model Elizabeth Couper-Kuhlen examines is that proposed by John Nerbonne in *German Temporal Semantics* (1984). While also favouring the principle of automatic progression, she argues against Nerbonne's proposal that the telicity of the propositions triggers progression of the reference time and suggests that a more detailed analysis of event notions is required.

The second model she considers is put forward in an article by Barbara Partee "Nominal and Temporal Anaphora" (1984). Consideration of Partee's discourse representations, which incorporate ideas from Hinrichs (1981) and Kamp (1979, 1984), lead her to conclude that, contrary to the currently accepted view, propositions representing states can move the reference time forward. She makes the interesting suggestion that only main clause events (i.e. foregrounded events) are responsible for the progression of reference times. Sub-clauses only trigger progression, she suggests, if they are narrative in character, that is, foregrounded. The progression rules as formulated by Nerbonne and Partee are, she concludes, in need of modification. The outcome of her critical examination of these models can be reduced to three essential problem areas which any model of temporal relations in discourse must address: (I) how to make temporal interpretation sensitive to changes in point of view which are not directly and unmistakably signalled in verbal form (e.g. certain forms of substitutionary speech and perception); (II) how to incorporate a notion of foreground and background based not only on syntactic form but also on discourse function (e.g. that of post-posed *when*-clauses), and (III) how to integrate the various principles of text construction and the various activities and strategies involved in text comprehension in a consistent and hierarchically ordered model of textual analysis.

In the second article, "The Awesome Seeds of Reference Time", Cornelia Hamann traces the development of Reichenbach's concept of reference time up to the present. She begins by pointing out a number of weaknesses in Reichenbach's tense analysis: (I) his mistaken conception of the English Present Perfect and the Progressive, (II) the unclassified status and function of reference time: whether it is to be conceived of as a time point or a time interval, whether it is a theoretical necessity for the analysis of tenses or a psychological entity, and whether it should be looked upon as a pragmatic or semantic concept; (III) the unclassified relationship between time adverbials and reference time.

Against attempts to dispense with the concept of reference time altogether, as in Prior (1967) and Bäuerle (1979), Hamann points out that this concept is necessary for (a) the analysis of tenses, above all for the distinction in English between the Past and Present Perfect tenses, (b) the understanding of the function of adverbs, (c) the distinction of the various event notions, and (d) the keeping of the time position in discourse. Regarding adverbials, she concludes that adverbially specified time and reference time must be seen as two distinct entities.

She then discusses the two attempts that try to take into account all of the above-mentioned points, namely Nerbonne (1984) and Partee (1984). The latter claims her particular attention since it is an attempt to develop "dynamic semantics" in the manner of Kamp's (1981) discourse representation as a method of representing the processing of temporally connected discourse. Finally, Hamann considers models of discourse representation using computer terminology and concludes her article by proposing three different kinds of activities or "commands" necessary for the analysis of tensed sentences in connected discourse, namely (a) commands establishing reference slots (as in the first sentence of narrations), (b) commands filling the slots, and (c) commands finding the slot, as in sentence sequences.

Hamann seems to advocate the notion of an automatic setting-up of the reference-time slot as soon as a sentence is tensed. The question remains however, as to how this slot is to be related to the reference time established by the preceding sentence in a temporally connected discourse.

A further perennial issue in our discussions was that of the relationship between time adverbials and reference time, a question left undiscussed by Reichenbach. In "Time Adverbials in English and Reference Time", Janet Harkness takes up this question.

The main concern of her article is first to distinguish between the time referred to by time adverbials (TA time) and reference time, speech time and event time, and then to indicate the connexions between these latter and TA time.

A distinction is first drawn between the times often contained in time adverbials (TA) and the times TA refer to, TA times. A further distinction is made between TA, which refer to times on the time line, and other temporal expressions, such as frequency and durative adverbials, which provide information about the events under consideration rather than the times.

TA time is taken as a fourth co-ordinate in temporal orientation; the foremost function of TA, it is argued, is to provide characterizations of reference time. A characterization can consist of an indication of the time span within which the reference time is to be posited or an identification of the time at or over which it is to be posited. If, for any of a number of reasons, the TA is not required to characterize reference time, it can characterize event time or even a time discrete from reference time, speech time and event time. The role TA play in combination with present perfect tenses differs somewhat from that in combination with simple past tenses; in combination with pluperfect tenses TA have characterizing functions parallel to those found with either present perfect tenses or with simple past tenses. Thus in present perfect tense sentences TA typically characterize event time as within or over a span of time beginning anterior to speech time/ reference time and extending up to speech time/ reference time. In simple past tense sentences TA typically characterize reference time and in pluperfect tense sentences they typically characterize either reference time or event time.

TA are not required in simple past tense sentences to indicate that reference is being made to a definite time; TA can, however, indicate that reference is being made to an unspecified definite time rather than to a specified. Thus *Harry was in Paris* is most appropriate in a context when the time referred to is established in some manner, whereas in *Harry was once in Paris*, the choice of this TA indicates that a specification or identification of the definite location of <Harry be in Paris> is not the issue at hand. A few TA indicate that indefinite temporal reference is involved (*some day, one day* (in future contexts)). With present perfect tenses the TA can, as appropriate, indicate that reference is being made to a definite time.

Our discussions saw us repeatedly confronted with the question of a functional analysis of English tenses. Of these, Present Perfect tense proved to be one of the most difficult to analyse and Richard Matthews' article "Present Perfect Tenses: Towards an Intergrated Functional Account" is an attempt to come to grips with the major issues involved.

He points out that accounts of the meaning of the English Present Perfect and Present Perfect Progressive are often contradictory. Some suggest a single basic meaning for these tense forms, others a number of (perhaps related) meanings, with labels such as 'perfect of experience', 'perfect of result', etc. At the root of this discrepancy lies the dichotomy between the meaning of an expression or construction (here the tense form) and its use, which includes its contextually determined implicatures. A brief survey is made of the various views, termed 'monosemic', 'disemic', 'trisemic', etc. which reveals a kind of hierarchically organized set of meanings and-or submeanings. In addition a cross-dialectal comparison made in Harris (1984), which seems to support the tetrasemic view, is subjected to re-examination. It is suggested that a monosemic view might in fact be tenable. Such a view of the present perfect tenses depends on a basic configuration in terms of speech time, reference time and event time in which the reference time is taken as terminating at speech time an interval within which an event or state holds. In order to put this to the test, two sets of predications reflecting various types of event notion (based on Schopf 1984 and Quirk et al. 1985) are examined both with and without temporal adverbial support. It becomes clear that the interpretation of perfecticized utterances can be almost exclusively predicted on the basis of the knowledge of the event notion, a basic temporal-aspectual meaning of Present Perfect and Present Perfect Progressive, and the temporal, aspectual or phasal meaning of adverbials. Discourse factors involved include: whether utterances are used to make descriptions or ascriptions, the thematic and rhematic function of adverbials, and the dependency relations of adverbial and tense: a) adverb-dependency; b) mutual dependency; c) tense-dependency. After a detailed discussion of the results, a set of hypotheses for the prediction of submeanings of the tense-forms and their collocation with adverbials is proposed, and an attempt is made to produce system fragments for the interaction of perfect tenses and event notion, and the combinatory meanings of tense and adverbial.

Alfred Schopf's article,"The Past Tense in English", puts forward three principal ideas: the first is that the nature of the reference time, i.e. whether it should be looked upon as a time point or time interval, is to some extent determined by the various event notions; states and unquantified or unbounded processes, for example, and activities (among other event types) tend strongly towards a pointlike reference time, while accomplishments require a reference interval. The second suggestion the article makes is that the Past Tense in English should be looked upon as an instruction to adopt or to look for a reference time, in most cases definite, in some marginal cases also indefinite. In contrast to other contributors to the volume,

Alfred Schopf seriously questions whether in temporally connected discourse the notion of an automatic progression of the reference time triggered by "event sentences" (i.e. sentences realizing certain event notions) can represent a realistic picture of the way narrative texts are processed by the listener or reader. Instead he suggests that the processing of narrative texts can be looked upon as a computational procedure in which the information supplied by every new sentence or clause is evaluated against the background of (all of the) information supplied by the preceding sentence and the larger context. Evaluation is made on the basis of semantic and pragmatic regularities at work in the language system in question and on the basis of general laws of cognition and our knowledge of extra-linguistic reality. Such considerations make him sceptical of the possibility of representing the decoding process for narrative texts in a completely formalized and mechanical manner.

The article also sets out to analyze the various speech and thought representation categories and provide a notational device for their representation. Substitutionary speech (or thought) and substitutionary perception, for instance, are looked upon as the interpenetration of two temporal universes, that of the narrator and that of the narrated character, each with its own sequence of reference times and events, the latter being embedded into the former by an embedding relation which correlates the narrator's reference time with the narrated character's speech time. Notational devices are also proposed for the representation of other, complex narrative techniques such as direct and indirect interior monologue. The article concludes by pointing to some peculiarities of English tenses.

The various approaches and ideas proposed in these five articles, as already pointed out, do not yet add up to a consistent, unified view of the temporal structure of narrative texts nor by any means to an exhaustive analysis of the tensing operations at work in the comprehension of text. Nor can we claim that all members of our research group were always convinced of every detail of a given analysis; we can, however, say that most were in agreement about most of the ideas put forward.

They do, however, provide what we hope is a useful and stimulating introduction to those questions being debated most heatedly in this fascinating and central area of linguistic research into English. The bibliography contains a representative selection of relevant work in this field.

TEMPORAL RELATIONS AND REFERENCE TIME IN NARRATIVE DISCOURSE

Elizabeth Couper-Kuhlen

1. On temporal relations in discourse

The aim of most linguists with an interest in time and temporal relations is to account for the *representation* in language of the temporal relations between two or more events in time or - seen from the recipient's point of view, to account for the *interpretation* of these same temporal relations, given their linguistic representation. In accord with the general trend of linguistics this century, scholars investigating time and temporal relations in language have proceeded along a path leading from the simple sentence to the complex sentence, and from there via the compound sentence to the text. It comes then as no surprise that the mainstay of current attempts to model the structure of temporal relations in discourse is a sentence-level approach to tense, namely that of Reichenbach. The aim of this paper is to explore how Reichenbachian tense semantics has been applied to temporal relations in discourse, in particular to narration, in two formal models and to evaluate these attempts critically based on a modest data base from the detective-story literature.[1]

1.1 Extending Reichenbach to discourse

The crux of Reichenbach's analysis of tense is the notion of *reference time* (see Hamann, this volume). For Reichenbach this is an abstract 'point' in time which by virtue of the way it configures with event time and speech time can be used to distinguish tense meanings in a given language. For instance, the configuration of point of event (E), point of speech (S) and point of reference (R) in the Simple Past tense in English, as in e.g. *Peter went*, looks as follows:

R,E - S

(where the comma indicates temporal coincidence and the dash, temporal precedence in a left-to-right reading)

---

1 Raymond Chandler, *The Big Sleep* (BS), Dashiell Hammett, *Red Harvest* (RH), *The Dain Curse* (DC), *The Thin Man* (TM), *The Maltese Falcon* (MF).

whereas for the Simple Past Perfect in English, e.g. *Peter had gone*, the configuration would be:

E - R - S (Reichenbach 1947:290).

Now, according to Reichenbach, the reference time within a simple sentence is frequently named by an adverbial. Thus in *I had seen him yesterday*, *yesterday* is understood as the point of reference anterior to which the event <I see him> occurred.[2] This is called the POSITIONAL USE of the reference point: the latter serves as "the carrier of the time position" for other temporal expressions. Positional use is also found, for instance, in compound sentences where "the time relation of the reference points compared is ... time sequence", as in e.g. *He telephoned before he came*. Here, Reichenbach claims, the R of <he telephone> is before the R of <he come>.[3] A special case of the positional use of the reference point is found in compound sentences such as *I had mailed the letter when John came and told me the news.* Here the R of <John come> is said to be identical with the R of <I mail the letter>: this Reichenbach calls PERMANENCE of the reference point.[4]

Now precisely these notions of position and permanence with respect to the reference point, developed primarily for use within simple and compound sentences, can be extended quite naturally to discourse. Schopf, for instance, distinguishes reference point CHAINING and reference point CLUSTERING in texts of more than one sentence, in particular in narrative texts (1984: 281ff.). With chaining he maintains: "each event and reference time coincide and the reference point last reached respectively forms an anchor for the temporal location of the next event" (1984:282; my translation). For example: *He went to town at 10; then bought a hat...* In this case the R of <he go to town>, which coincides with E due to the Past, serves as an anchor for the E of <(he) buy a hat>.[5] With clustering, on the other hand, "different

---

2 In a somewhat different interpretation *yesterday* could of course also specify the time when the event <I see him> took place.

3 Cf., however, Schopf (1984), who maintains that only one R is necessary for the temporal interpretation of this sentence.

4 Note, however, that if this *when*-clause is interpreted as a 'narrative' clause, i.e. one which carries the foregrounded events ahead in time, then its R must be later than that of <I mail the letter> (cf. Couper-Kuhlen, forthcoming).

5 This holds, however, only if <he go to town> is interpreted punctually. If it involves a process (in which case it would presumably belong to Vendler's Accomplishment category), then *at 10* specifies only the time of the commencement (t_i) of the process phase: <(he) buy a hat>, however, must be anchored to the completion (t_f) of the whole event (see also Hamann and Harkness, this volume).

events are related to one central reference point without themselves having one" (1984:286; my translation), as in e.g. *I arrived at Aunt Mary's at 12. In the meantime John had called Aunt Mary to persuade her not to see me.* Here, it is assumed, the event <John call Aunt Mary> is related to the R of <I arrive at Aunt Mary's>,[6] which itself is specified by the time adverbial *at 12.*

1.2 The need for reference time in discourse

Although it has been claimed in the past that the notion of reference time is redundant in the analysis of Past tense,[7] recent arguments tend to emphasize, on the contrary, that reference time is indispensable at the level of discourse, at least for certain kinds of discourse (cf. e.g. Nerbonne 1982, Reinhart 1984). For instance, it is pointed out that the 'truth' or coherence of a narrative such as *John washed his face. He went downstairs* is not fully constrained by simply stating that due to the Past tense E and R coincide and precede S in each clause. This imposes no restrictions on the relation between the E's, R's and S's respectively of the two clauses. If the events did indeed take place in the order mentioned, then the narrative is only an accurate representation of them on the condition that E,R and S of clause 2 follow E, R and S of clause 1. If they do not, then the discourse is 'faulty' in some important sense.

This argumentation can be extended to 'multi-tense' narratives, e.g. *John had washed his face. He (now) went downstairs.* Here it is not sufficient merely to state E-R-S for clause 1 and E,R-S for clause 2 without stipulating that E_2 follows E_1 ($E_2>E_1$) and that R_2 coincides with or follows R_1 ($R_2 \geq R_1$). And for *John washed his face. He had gone downstairs (to do so because that was where the only hot tap was)* it must be stipulated that E_2 precedes E_1 ($E_2<E_1$) and that R_2 is the same as R_1 ($R_2=R_1$). What appears to characterize these various narratives is not the relation between the E's in linear succession - this may vary - but the relation between the successive R's[8]: in all cases the R of a following sentence either remains the same as or follows on the R of the preceding sentence.[9] Thus to account for the temporal coherence of this kind of discourse we must assume a constant relation between the R's of consecutive sentences - which amounts to an indirect proof of the necessity for R at the

6 The Past Perfect signalling that it is anterior to this R.
7 Cf., for instance, Bäuerle(1979:49).
8 And trivially, between the successive S's.
9 These are of course the clustering and the chaining principles respectively.

level of discourse.

Note that the restrictions placed upon successive R's to cluster or chain is not true of <u>all</u> types of discourse but only of those types which have been called (somewhat circularly) 'temporally connected' (Nerbonne 1984). Thus *Al went to N.Y. The others were there once too* is perfectly coherent as a text, although R_2 is not necessarily the same as or later than R_1. (Indeed the event <the others be there once too> is not related temporally in any specific way to <Al go to N.Y.>: it could easily be true before, after or during the time when <Al go to N.Y.> is true.) Likewise a text originating in answer to the question 'What happened yesterday?' such as *Bill went to the movies. Jane wrote a letter. John dug up the garden* (Kiefer, to appear) is well formed although the R's need not succeed one another in the order given. In this case, as in that above, the temporal relation of the events to one another is irrelevant to the textual strategy employed. The type of temporally connected discourse most commonly cited as observing the R-chaining/clustering principle is *narration*, although it is by no means the only type. 'Texts' such as *The company announced that supplies were running low* or *They said on Sunday that they were leaving in 3 days* also satisfy the definition of temporal connection.

2. Two Reichenbachian models of temporal relations in discourse

We turn now to an examination of two formal models based on Reichenbachian tense semantics.

2.1 An indexical model

Nerbonne (1982, 1984) treats Reichenbach's reference time, speech time and event time as contextual parameters associated with tense: in model theoretic terms they become indices t_r, t_s, t_e. The truth of a sentence in a given tense is determined with respect to the time values of these three indices. For a sentence in the Past, for instance:

$$A_{s,e,r} \models \text{PRET}\ (p) \text{ iff } e = r < s \text{ and } A_{s,e,r} \models p \qquad (1984:21)$$

That is, a sentence *p* in the preterite is true in model A relative to *s,e,r* if and only if event time equals reference time and both precede speech time and *p* is true in model A relative to *s, e, r*.

2.1.1 Reichenbach's Pragmatics

Now since, as we have seen, it is necessary to specify not only the proper configuration of these three times for each tensed proposition but also the relation between these times respectively for each successive proposition, Nerbonne introduces the following pragmatic principle:

> Reichenbach's Pragmatics (RP) (weak version): For $S_1, S_2, \ldots, S_n$ a sequence of sentences uttered in a temporally connected discourse:
>
> (i) $r(S_i) \leq r(S_{i+1})$
>
> where r(S) designates the reference time of S. (1984:8)

(The symbol /$\leq$/ is to be read as "does not extend beyond".) What this principle says in effect is that the R of a given sentence in temporally connected discourse does not extend beyond the R of the following sentence. Since Nerbonne follows Bennett/Partee (1978) in assuming that time *intervals* rather than points are primitve, this means that the following reference time configurations would NOT be possible:

(i) 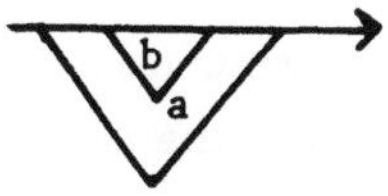

(ii) 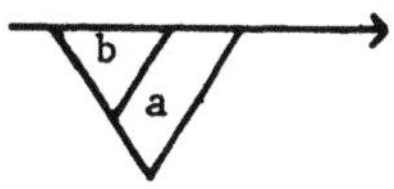

(iii) 

(iv) 

In all of these, the reference time of the event in a first sentence (a) extends beyond that of the event in a following sentence (b). With respect to a narrative such as *Al went to N.Y. The others were there too*, this would mean that the R of <Al go to N.Y.> extends beyond that of <the others be there>. Now since Nerbonne assumes with Reichenbach that when the Past tense is used in English, reference time coincides with event time, the above configurations would correspond to the following event-time situations: (i) Al's going to N.Y. begins before and extends beyond the others' being there, (ii) the others' being in N.Y. begins simultaneous with Al's beginning to go to N.Y. but terminates before he reaches his destination, (iii) the others leave N.Y. immediately before Al begins to go there, (iv) the others begin to be in N.Y. before Al begins to go there but leave N.Y. before he arrives. These situations are of course not impossible, but discourse which describes them as above would be 'temporally free' in Nerbonne's terms. (Note that Nerbonne is obviously considering only cases of straightforward narration with a 'visible' narrator; a reading of *The others were there too* as free indirect discourse would presumably permit both (ii) and (iv) and the discourse above would still be temporally connected.)

Interpretations which ARE permitted by *Reichenbach's Pragmatics* (weak version) include:

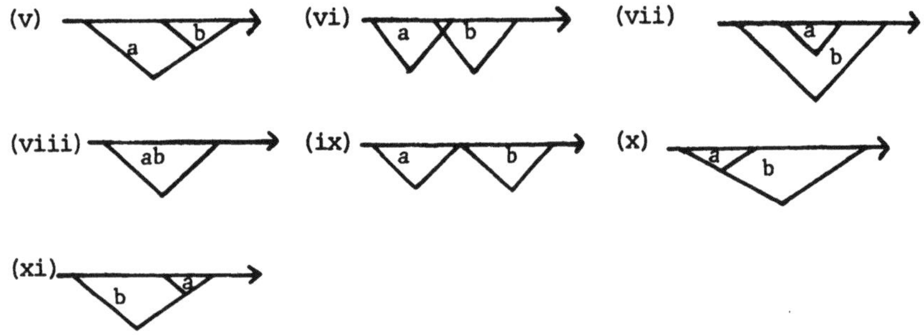

In none of these configurations does the reference time of event (a), say <Al go to N.Y.>, extend beyond the reference time of (b), say <the others be there> - if the notion of extending beyond is understood to apply only to the *right*-hand border of the reference time interval, as Nerbonne apparently intends it to do. This corresponds respectively to situations in which (v) the others begin to be in New York after Al begins to go there and leave simultaneous with Al's reaching his destination, (vi) the others begin to be in N.Y. after Al begins to go there and they continue to be there after Al's arrival in N.Y., (vii) the others begin to be in N.Y. before Al goes to N.Y. and they remain there after he arrives, (viii) the others begin to be in N.Y. at the same time as Al begins to go there and they cease being there at the same time he reaches N.Y., (ix) the others begin to be in N.Y. immediately after Al reaches N.Y., (x) the others begin to be in N.Y. at the same time as Al begins to go to N.Y. and they remain there after he arrives, and (xi) the others begin to be in N.Y. before Al begins to go to N.Y. and they leave at the same time as he arrives.

Note, however, that the weak version of *Reichenbach's Pragmatics* also permits:

(xii) 

In this case there is an unspecified gap of time between the R of (a) and that of (b). This would correspond to an event-time situation in which the others don't begin to be in N.Y. until some time after Al arrives there. According to Nerbonne, this interpretation should be prohibited - or rather, only permitted if a conversational implicature has been cancelled. We must therefore strengthen *Reichenbach's Pragmatics* in such a way that (ix) becomes a conversational implicature. Nerbonne apparently assumes that the problem is related to the 'aspectual class' or event-notion type of the second predicate, which in Vendler's (1957) terms is a State here, not an Activity, Achievement or Accomplishment,[10] and proposes to modify the pragmatic principle as follows:

<u>Reichenbach's Pragmatics</u> (strong version): For S_1, S_2, ... S_n a sequence of sentences uttered in a temporally connected discourse:

(1) $r(S_i) \leq r(S_{i+1})$

(2) it is conversationally implicated (for all atelic S_i) that $r(S_i) \leq r(S_{i-1})$

where S is atelic iff it is an activity or a state. (1984:23)

The rider (2) here ensures that the reference time of a State or Activity does not extend beyond that of the preceding event. Now, since (1) ensures that the R of an event does not extend beyond that of the following event, the upshot of the revised version is that the R's of States and Activities must end simultaneously with the R of the preceding event. That is, only configurations (v), (viii) and (xi) are possible if (b) is 'atelic'.

2.1.2 The strong version's weakness

But the effect of the strong version is unfortunately to throw the baby out with the bath water. That is, it not only excludes an undesirable configuration of R's but also a number of desirable ones, namely (vi), (vii), (ix)

---

10 In the following we shall distinguish event-notion categories such as State, Activity, Accomplishment, Achievement from generic events or eventualities by the use of initial capitals.

and (x), whose occurrence can be readily documented. Consider, for instance:

(1) I opened the refrigerator door. It felt cold inside.

(2) We went into her cluttered-up living room. Dan Rolff and Max Thaler were playing pinochle there. (RH62)

(3) James switched off the light. It was pitch dark around him. (Hinrichs 1981:66)

(4) She tossed her head angrily and the rich color of it glistened in the dim light. (BS 9)

The event-time configurations in examples (1) and (2) correspond presumably to (vi) and (vii), in examples (3) and (4) to (ix) and (x) respectively. That is, in all these examples the State/Activity predicated in the (b) sentence extends beyond that of the preceding event. And since in Reichenbach's and Nerbonne's tense semantics event time coincides with reference time in the English Past, we must conclude that the second R also extends beyond the first. Yet this is precisely the temporal relation between Event and State/Activity which Nerbonne's strong principle excludes.

The problem here can be traced to Reichenbach's view that R is equivalent to, i.e. *coincides* fully with event time for all verbs in the Past irrespective of 'aspectual class'. There is a way out of the dilemma created by States and Activities (the proposal comes originally from Dowty), which Nerbonne sketches as follows:

> States and activities may be true of times, without for that reason being false of their subintervals. Thus if *Jones be at home* is true of 9-10 am, it isn't therefore false of 9:00-9:10 am. If this sort of sentence is S_i within a connected discourse, one may suppose it to be predicated of an $r(S_i)$ such that $r(S_i) \leq r(S_{i-1})$ -- even if it is also true of superinterval $j\ r(S_i)$ such that $j \,-\!\leq\, r(S_{i-1})$. (sic)* (1984:24)

(/-≤/ is to be read as 'does extend beyond'.)

What this amounts to is a new understanding of reference time as a time which is included in event time but need not necessarily be wholly equivalent to it. If Jones is at home from 9-10 am, this is event time. If we check to see whether <Jones be at home> is true at 9:10 am, then this is reference time, the time at or during which we evaluate the proposition for its truth. Likewise, if <John be building a house> obtains from January-October approximately but we check to see whether it is true on April 4, then the latter interval is reference time. This modification of the notion of reference time, it can be

---

* Presumably there is an inclusion sign missing which should link j and $r(S_i)$ at the beginning of this line: $j \supset r(S_i)$.

argued, is necessary for the interpretation of states and progressives, which are characterized by the fact that they have no t_i/t_f or respectively that their t_i/t_f are phased out (cf. Schopf 1984). Thus their event time can never be got at fully; it is only via reference time that we can evaluate their truth or falsity. If the truth of a state or a progressive in the Past depends on R being included in event time rather than fully coinciding with it, then the problem discussed above disappears.

However, it is significant that in attempting to constrain *Reichenbach's Pragmatics*, Nerbonne does not fully succeed in his original purpose, which was to block (xii). This configuration is still possible if both sentences are telic.

2.1.3 The weak version's weakness

Is the weak version of *Reichenbach's Pragmatics* of any greater advantage? Rather than prescribing one (or several specific) temporal interpretation(s) for a given event-notion configuration, the weak version appears to allow any interpretation for any combination of Event, State and Activity. The problem with such an 'anything goes' model, however, is that event-notion combinations do constrain the options for the interpretation of the temporal relations in narrative discourse in predictable ways. For instance, in the following excerpt from a detective story:

> (5) Dan Rolff was in the living room. A water glass and a brown bottle with no label stood on the table in front of him. He sat straight up in his chair, staring at the bottle. (RH 75)

the event <he sit straight up in his chair> is open to two different temporal interpretations, depending on whether we treat it as an Activity with no clear initial or ending point, in which case it would be approximately equivalent in meaning to 'he was sitting straight up in his chair', or as an Accomplishment with a clear beginning point, equivalent in meaning to 'he began to sit straight up in his chair'. In the former case the event time (and presumably the reference time) coincides with that of the preceding State <Dan Rolff be in the dining room>. In the latter case, however, the event time (and reference time) of the Accomplishment intervenes during that of the preceding State.

The upshot is that in the first case there is no moving forward in (narrative) time, whereas in the second case the story is clearly propelled forward by the action which intervenes. Thus depending on event-notion category the temporal interpretation of the events in this passage is either one of overlap between two unbounded intervals or of a clearly bounded interval intervening during an unbounded interval - and in fact this can be shown to be systematically so depending on event-notion configuration. Yet *Reichenbach's Pragmatics* (weak version) does not constrain the relation between successive E's and R's as a function of event-notion and so is grossly underspecified. It may not be 'wrong' if it is regarded simply as the collapsed version of a number of more specific rules, but it is of little help in predicting which interpretations are possible given a particular event-notion combination.

2.1.4 Telicity as a sequencing criterion?

Even when Nerbonne does attempt to formulate constraints on temporal interpretation as a function of event notion, viz. in the strong version of *Reichenbach's Pragmatics* (Nerbonne 1984), however, the categories adopted are not the pertinent ones. According to Nerbonne, the crucial factor in determining whether an R may extend beyond the preceding one is telicity. Cf.:

> (2) Smith walked around. Jones was at home.
> Brown talked with Jones.
> (3) Smith came in. Brown wrote a letter.
> Brown arrived.
> I.e., the normal understanding of atelic following atelic is that they are nearly simultaneous (2), while the normal understanding of telic following telic is that the first precedes the second (3). (1984:24)

But if *telic* is being used in its usual acceptation, namely to denote verbs/predicates which "involve() a *process* that leads up to a *well-defined terminal point*, beyond which the process cannot occur" (Comrie 1976:45; my italics), then telicity is too restrictive a factor to be responsible for the sequencing (vs. simultaneity) of events. Indeed certain non-telic predicates will also be understood to refer to events in succession in narration:

(6) The balloon popped. The child jumped.
(7) The guide looked up at the sky. The tourists looked up at the sky.

Nerbonne is thus right in according event notion the role he does in the interpretation of temporal relations in temporally connected discourse but he is misled in making telicity the criterial feature.[11]

2.2 A discourse representation model

The second formal model to be considered here also recognizes the importance of event notions for temporal interpretation. However, it employs different categories (basically Vendler's) and, more significantly, incorporates them explicitly into a quite different formal semantic framework. The tenets of discourse representation can be traced back to Kamp (1979, 1984); temporal models in this tradition have been worked out by Kamp/Rohrer (1983) for French and by Hinrichs (1981) and Partee (1984) for English. The following discussion deals primarily with Partee (1984).

A discourse representation model emphasizes above all the processes a discourse recipient goes through on a sentence-to-sentence (or utterance-to-utterance) basis in the course of constructing discourse meaning:

> ...consider a very simple situation, in which one person, S, does all the talking, and another, H, all the listening. (...) As the discourse proceeds H will represent the information conveyed to him; and we shall suppose (...) that he modifies his representation after each successive sentence that S utters. Thus H produces a series of representations D_1, ..., D_n. Each next representation D_{i+1} results from incorporating the new information contained in S_{i+1} into the structure D_i which represents all the information obtained from $S_1,\ldots,S_i$. (Kamp 1979:381)

Tense in such a model can be thought of as a set of directions to construct a particular representation of times or rather, since it is customary to take events rather than instants as primitive in this approach, a particular representation of *events*, entities of finite duration[12], which correspond to the Reichenbachian parameters E, R and S.

2.2.1 Event notion and *r*

For the representation of temporal meaning in narrative discourse, Partee assumes that the first sentence of a narrative has a past reference time (*r*) specified in some way at the outset. Thus we begin with r_1 which is prior to *s*. The event of the first sentence is then located with respect to this *r*

11 The event-notion parameter which is more directly linked to sequential vs. simultaneous interpretation of events is, we claim elsewhere, initial determination (Couper-Kuhlen, in preparation).

12 In the case of R, descriptively 'empty'.

according to the following principles:

(i) If it is an *Event*, i.e one of Vendler's Achievements or Accomplishments, it is placed somewhere within the current *r*, in other words the reference time <u>includes</u> the event time:

$e_i \subseteq r_o$
where r_o is the current reference time and $/\subseteq/$ represents 'is included in'.[13]

This corresponds intuitively to our understanding of what events are, viz. situations which are, or are seen as unitary wholes with definite beginnings and endings (Galton 1984:24ff.) and which occur within intervals rather than at single moments:

> ...since an event in general takes time, it cannot ever be wholly present at one time; while a state, although it may endure over a stretch of time, does not change during such a stretch and so is present at each moment of the stretch. (Galton 1984:15)

Thus when we place an event within the current *r*, we are in essence saying that we cannot necessarily single out a <u>moment</u> when this event is true but we can meaningfully predicate this event of an interval included within a larger interval.

(ii) If it is a *State* it is located around the current *r*, or in other words the reference time <u>is included in</u> the event time:[14]

$r_o \subseteq s_i$.

This too seems intuitively correct given what we know about States, namely that they obtain over an interval and indeed do so at every moment within that

13 Note that this use of the term *inclusion* encompasses complete overlap so that the following relations between event time and reference time are possible:

14 Here too complete overlap is possible, so that we obtain as possible configurations:

interval:

> ...states are primarily to be associated with moments, and events with intervals. A state may obtain in an interval only by virtue of its obtaining at some or all moments of the interval, whereas an event is first and foremost to be located within an interval. (Galton 1984:24f.)

Since *r*'s in Partee's model are thought of as Events, it makes sense for them to be located within the stretch of time established by a State. We can single out an interval at which we can verify the truth of a State, but the State itself may extend beyond the borders of this interval. The reference time of a State is thus only a subinterval of its event time.

2.2.2 *r*-progression

The second assumption that Partee makes is that "the introduction of new event sentences moves the reference time forward" (1984:254). In other words, with every successive introduction of a new Event into the universe of discourse, provided this Event is introduced via the predicate of a main clause, a new reference time is automatically introduced:

$$e_i \lesssim r_{i+1}$$

where / $\lesssim$ / represents precedence with assumed proximity

i.e., introduce a new *r* immediately after the Event last processed. This new *r* becomes the current reference time to which the event in the next sentence is related according to the event-notion principles outlined above. Note that it is only Events which cause *r* to shift ahead automatically. If the eventuality being located is a State, the current *r* remains where it is. This has the consequence that any eventuality which *follows* a State must be located with respect to the same current *r*. Successive eventualities accumulate, so to speak, with respect to this *r* until the next Event comes along, which introduces a new *r*.

2.2.3 Adverbials and *r*-progression

When *r* is shifted ahead and the next sentence contains an adverbial - say, *at six o'clock, three years later,* or *after an hour* - the current *r*, according to Partee, is 'updated', i.e. presumably it is readjusted so as to correspond to the reference time specified by the time adverbial in the next sentence. (For problems associated with this, see Hamann and Harkness, both in this volume.)

In the case of a temporal clause in the next sentence - say, with *when* - the subordinate predicate introduces a new *r* of its own, according to Partee: "a (preposed) *when*-clause in a linear narrative triggers the introduction of a new reference time located 'just after' the event described in the *when*-clause" (1984:261). Similarly, a *before*-clause introduces an *r* which is located before its event and an *after*-clause, one which is located after (but not necessarily just after) its event. The temporal-clause event itself, however, "is not constrained to occur within the then-current reference time, but rather serves to provide a descriptive anchor for the next-introduced reference time" (1984:261f.).[15] It apparently is constrained, however, to introduce a new *r* which comes after the then-current *r* by virtue of the added stipulation that in a linear narrative mode "successively introduced reference times strictly follow one another" (1984:260).[16]

2.2.4 Strict succession of *r*'s?

In addition to the problems associated with Partee's treatment of temporal clauses (see also Hamann, this volume), the stipulation that reference times strictly follow one another in linear narrative is at best contradictory when combined with the progression rule. According to the latter, a new *r* is introduced just after the last processed Event, i.e. diagrammatically:

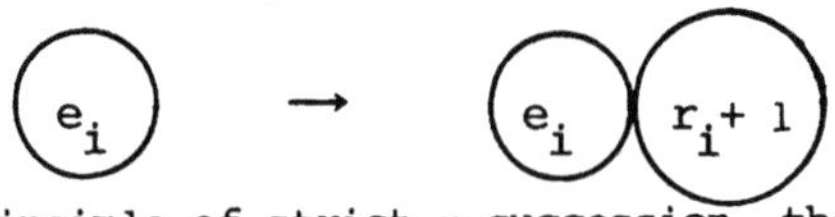

But according to the principle of strict *r* succession, the *r*'s must be contiguous, i.e.

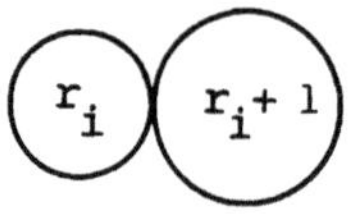

It should be obvious that these two principles can only be reconciled if *e* and *r* have their right-hand border in common:

15 In fact, in all of Partee's diagrams the event of the temporal clause is located to the right of, i.e. <u>after</u> the then-current *r*.

16 Partee apparently overlooks the fact that the reference times established by *when* and *after* do not necessarily follow strictly on the then-current *r*, since the temporal-clause event may come inbetween.

However, if e_i is included in r_i such that

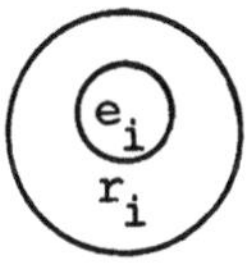

is the next *r* positioned immediately after e_i or immediately after r_i? If the former, then the *r*'s do not strictly follow one another. If the latter, then the progression rule as it stands needs revision.

3. Three problems for models of temporal structure in discourse

In this section I shall not deal further with specific sins of commission or omission in the two models at hand. Instead I shall discuss three problems which I believe any model (and by inclusion, these two specific models) of temporal structure in discourse must address. These relate to the role of (i) reference time, (ii) syntactic form and (iii) tense, all keystones in the structure of current models.

3.1 Reference time and point of view

The formal models under consideration - and indeed probably any formal model - must of necessity take tense in the text at face value, so to speak. That is, since it is the narrator who is telling the story in his/her own words, we assume, the reference times of the Past tense in narration are the narrator's. Subtle changes in point of view - to the extent that they are not marked in the verb - will consequently go unobserved in the model and temporal misinterpretation may result. Consider, for instance:

(8) I sipped my drink and nodded. The pulse in his lean grey throat throbbed visibly and yet so slowly that it was hardly a pulse at all. An old man two-thirds dead and still determined to believe he could take it.
'Your conclusions?' he asked suddenly. (BS 14)

At first glance it might appear that the R of the Activity <the pulse in his ... throat throb> and the R of the State <(he be) an old man two-thirds dead...> should overlap or coincide with the preceding Event <I nod> - or in an automatic progression model, with the current *r* established by this Event. But this would have as a consequence that no (narrative) time elapses between <I nod> and <he ask>. The Activity and the State would simply be treated as simultaneous to some previous Event, not in any sense Events themselves which propel the story forward. And yet there are cues in the text[17] - albeit unrelated to tense - to the effect that the Activity <his throat ...throb> and the State <(he be) an old man> should be interpreted as substitutionary perception (Fehr 1938) or free indirect thought (Leech/Short 1981), the result of a conscious act of observation or reflection on the part of the narrator, which a 'visible' narrator would report with 'I noticed...' or 'I thought...'. Here, however, the narrator (the narrating-I) has slipped into the consciousness of a character (the narrated-I) and uses the latter's words and thoughts to propel the story forward. Unless a formal model can be made as sensitive to switches in point of view as the astute reader, it will always present an oversimplified, 'flat' view of the narrative world.

3.2 Foregrounding: syntactic form or discourse function?

Events which constitute the backbone of a story, i.e. those which push the narrative forward via R-chaining or *r*-progression, it is assumed in current models, are main-clause events. Events located in subordinate clauses, it is claimed, are part of the background; they locate the main-clause events by helping to specify an R for them. This view of locating (R-specifying) and located (R-specified) events, however, is heavily dependent on a view of syntax which is traditional and sentence-bound. Recent discourse-oriented work (e.g. Harris 1986, Matthiessen/Thompson (forthcoming), Harris (forthcoming)) has shown that the syntactic opposition main vs. subordinate clause does not always correspond to the discourse functions of 'nucleus' vs. 'satellite', to use Matthiessen/Thompson's terminology. If we understand by 'nucleus' a clause which functions to create the *foreground* in narrative discourse, then this view is corroborated by the existence of so-called *narrative temporal clauses* (Couper-Kuhlen (forthcoming)), i.e. foregrounded *when*-clauses which carry the story line forward. Consider, for instance:

---

17 E.g. *visibly* and the elliptic construction of the third sentence.

(9) I didn't move. She gave me another cute glance and went on towards the front door. She had her hand on the knob when we both heard a car coming. She looked at me with questions in her eyes. I shrugged. The car stopped, right in front of the house. (BS 59-60)

In this context it is quite clear that the foreground consists of the following events: <she give me another cute glance><(she) go on towards the front door><we both hear a car coming><she look at me><I shrug><the car stop>. In other words, the event <we both hear a car coming> is an integral part of the story line, although it is in a subordinate clause. Instead of the temporal-clause event helping to specify the reference time for the main-clause event, the functional roles of main and subordinate clause are reversed here: it is the main-clause event with respect to which the temporal-clause event is located. Therefore models such as those examined here, which determine the interpretative rôle of a temporal event based solely on a syntactic distinction between main vs. subordinate-clause, will automatically overlook narrative temporal clauses of this sort. The faulty temporal interpretations which may ensue from such oversight have been discussed elsewhere (Couper-Kuhlen (forthcoming)). The point to be emphasized here is that until formal models can be made to orient to discourse function (i.e. foreground vs. background in narration) instead of sentence-level syntax, they will be inadequate both as 'static' representations of temporal structure and as dynamic models of the interpretative process.

3.3 Tense and text type

The two models under consideration, and presumably any model of temporal relations in discourse constructed on similar principles, set up rules of interpretation which depend crucially on choice of tense. It is the presence of Past tense, for instance, which triggers the configuration of R, E and S, or modified versions thereof, upon which the interpretation of temporally connected or narrative discourse rests. The implications of this are threefold: (i) conditions for the interpretation of narrative discourse are the same as those for the interpretation of temporally connected discourse; (ii) narrative discourse occurs only in the Past; (iii) there is a free choice of tense in every sentence of narrative discourse. But none of these assumptions is wholly justified.

Note first of all that the notion of 'temporally connected discourse', as we have seen, is much broader in scope than that of narration. Narrative texts, to quote one widely cited definition, must involve at least two temporally ordered events which are reported in so-called narrative clauses, i.e. main clauses the linear order of which preserves 'natural' order (Labov/Waletzky 1967, Labov 1972). According to this definition narrative entails at least two main clauses and two sequential events reported in iconic order and is consequently a more restricted subset of temporally connected discourse. A model which purports to account for temporally connected discourse is therefore likely to be underspecified for the description of temporal relations in narration since the latter are even more constrained.

However, even when models purport to account only for temporal relations in narrative, to do so via the Past tense is somewhat shortsighted. Based on the definition of narrative above, this kind of temporal organization occurs not only in the Past but in the Present and the Future.[18] In fact it can even be spotted in the Past Perfect, as the following excerpt illustrates:

(10) While we were talking about it, a pair of plain-clothes men brought in the red-faced bird who had stopped the slug I had missed Whisper with. It *had broken* a rib for him, and he *had taken* a back-door sneak while the rest of us were busy. Noonan's men *had picked* him *up* in a doctor's office. (RH 102)

Consequently the rules for locating eventualities with respect to current reference times which are established via a chain of successive Events are equally as applicable to narrative in other tenses and must be repeated quasi verbatim in the rules for the interpretation of other tenses, when they are used in narration.[19] This suggests that for the organization of temporal relations in narration the contribution of the Past tense (as opposed to some other tense) is minimal,[20] and that what we are actually

---

18 See Chafe's (1980) Pear Stories for examples of (oral) narration in the Present, and Bronzwaer (1970:70ff.) for an example of narration in the Future.

19 For an example of this type of redundancy, cf. Hinrichs' rules for interpreting Past and Past Perfect.

20 Cf. also Dahl (1984); Comrie (1986) comes to the same conclusion not only for 'single-tense' but also for 'multi-tense' narration.

dealing with is a particular text type - e.g. narration - with a characteristic strategy of text composition which consists in organizing events this way. (For the notions of text type and text strategy, cf. Enkvist 1981.)

A second indication that the contribution of tense in narration is minimal as far as temporal interpretation is concerned is the fact that under certain conditions Present can be substituted for it (cf. e.g. Wolfson 1979), just as under certain conditions Past can substitute for Past Perfect in narrative sequences. Cf.:

(11) McGraw told me he *had sent* a pair of his dicks (...) to see the girl that morning, to see how much help she could and would give the department in copping Whisper for Noonan's murder. The dicks *got* to her house at nine-thirty. The front door *was* ajar. Nobody *answered* their ringing. They *went in* and *found* the girl lying on her back in the dining-room, dead, with a stab wound in her left breast. (RH 157)

In both cases it is usual to speak of a 'neutralization' of the distinction between tenses which otherwise contrast in similar or identical environments. However, it may be that the simple transferral of sentence-level tense contrasts to the analysis of tense in connected discourse is a mistaken effort in the first place. It allows for no global text strategies and makes the occurrence, say, of a Past tense form in the first sentence of a discourse just as 'informative' as one several sentences later. Yet as is becoming apparent from universal tense research, some languages have a special *narrative tense* which is used consistently once the first sentence of a discourse has established that it is narrative (cf. Dahl 1984). This implies that the first tense choice in a discourse is maximally informative and that tense in following sentences in narration is predictable and of little informational value, or rather only of informative value when the expectations set up by text type and strategy are not met. Sentence-level tense semantics carried over on a sentence-by-sentence basis to narrative discourse does not do justice to such text semantic macro-structures.

Ultimately this is the lesson to be learned from the problems which current models of temporal relations in discourse encounter: discourse cannot be treated as a string of sentences organized at some local level only, nor can temporal interpretation in texts be reduced to a reiterative application of syntax-driven tense semantic rules.

THE AWESOME SEEDS OF REFERENCE TIME

Cornelia Hamann

0. Introduction

As the concept of reference time plays a role in the work of most authors concerned with the analysis of temporal structures and proved to touch on almost any field the TSE-Project investigated, this is an attempt to outline its origins, its favourable or unfavourable reception and its development.

Reference time turns out to be relevant for the analysis of tenses, adverbs, event notion and temporal text-cohesion, so that, quite predictably, the problems grow proportionally to the degree of integration of those four fields.

The article, it is hoped, will clarify the concept and also its problems, it does not purport to present solutions or a unifying theory. Solutions to some of the problems are sketched in the other contributions to this volume. The article grew out of discussions of the problems each of the project members came across in his/her work. Therefore it contains many arguments that became common property and cannot be traced back to one person alone. In these discussions there also grew a general conviction that reference time in a modified Reichenbachian sense is an indispensable concept for any description of temporal structures.

Part I Planting the Seed

1.0 Reichenbach

It was the difference in meaning in the sentences

(1.1) Tom arrived yesterday
(1.2) Tom had arrived yesterday

that led grammarians[1] and philosophers to postulate not only the point of

---

1 Cf. Jespersen (1931:60 ff.).

speech and the time of the event as the relevant points for the interpretation of tenses, but also a third time, the so called reference point or reference time.

Since Reichenbach (1947) was the first to employ this concept in a systematic description of natural language tense systems, it is only fair to begin our investigation of the erratic fate of the reference point with an account of his original system.

1.1 Reichenbach's three points analysing tense and aspect

Reichenbach's description of the tenses starts with the observation that tenses are token-reflexive since "they determine time with reference to the time point of the act of speech, i.e., of the token uttered" (Reichenbach 1947:288).[2] Therefore at least two points are involved in any adequate description of tense: the time of the event (event time, E, ET) and the time of speech (speech time, S, ST).[3] Those two times give rise to three basic time relations: "before the point of speech", "simultaneous with the point of speech" and "after the point of speech".

This formulation shows three things:

(a) The point of speech serves much the same purpose as a constant in a model of formal logic,[4]

(b) The constant divides the time line into three major domains: the now, the past and the future, and

(c) The event is being located with respect to the point of speech.

From here the line of argument is obvious: If only the two above mentioned points were involved, we would have only those three relationships and therefore only three tenses.[5] As most languages have more than three tenses, something else must be involved. In view of sentence (1.2) this "something

2 Tichŷ (1980:354) challenges this view especially for the Simple Past. He analyses a sentence in the Simple Past in such a way as to make it entirely independent of the time at which it is uttered.

3 E and S are the abbreviations Reichenbach uses for event time and speech time. They will be employed henceforth in the strict Reichenbachian sense. When we are speaking more loosely or metalinguistically and after we have agreed to modify the concepts especially that of reference time (R in Reichenbach) we will use ET and RT or ST as abbreviations.

4 As an example might serve a dense order with a constant. If we interpret the constant as 0, we have a model in which negative and positive numbers can be quite easily defined. In a way, we can say more about the elements of our structure than in a model without a constant.

5 Turning the argument around, we could say that for the Simple Tenses, three in number, we therefore should not need anything else but RT and ST.

else" is, according to Reichenbach, another event, a point in between the time of speech and the time of Tom's arrival.[6] This third point is called the "point of reference".

For an isolated sentence like (1.2) it is not obvious which of the available points between E and S is being used as a reference point. The tense only indicates that there is one such point, and it is the context (of speech) which determines it.[7] Reichenbach's example is a passage from Maugham where a series of events are recounted in the Simple Past and the use of a Past Perfect then refers the event in the Past Perfect sentence not directly to the point of speech, but to one of the events in the story. To illustrate the point the following may serve:

(1.3) a) He came home late. b) He stepped into the living-room and felt that something was wrong. c) He had put the car into the garage as usual and had walked up to the front door whistling softly. d) Nothing had stirred. e) Now he tensed.[8]

Instead of this implicit determination by context we can have an adverb, which determines the reference point explicitly:

(1.4) a) No, you can't have seen him at three thirty. b) I made inquiries. c) At three o'clock he had already left the building.

Here it is the adverbial *at three o'clock* that determines the reference point for the Past Perfect in (1.4.c).

1.1.1 The six English tenses

There seems to be little doubt and controversy about the existence of an implicit or explicit past reference point as long as we are dealing with a Past Perfect. But Reichenbach went further: He postulated such a reference point for all tenses, even for Simple Past, Simple Present and what Reichenbach calls Simple Future, which at first glance seem to involve only speech and event time. Reichenbach, however, considers the possibility that two of the three points can coincide. Almost in passing he then produces a brilliant distinction

---

6 It is interesting that Reichenbach speaks of an event here. He might simply be saying that some events serve as reference time, or he might be using event as the basic concept from which intervals and points can be defined (cf. Wiener (1914)).

7 Reichenbach says "determination" of the point (1947:288).

8 I depart from Reichenbach because the Maugham text involves the verb *think* and the rules of Reported Thought are rather involved. But even my mini-story had to be elaborate. I needed (1.3.b), because otherwise putting the car away could be construed as a subevent of the complex event of coming home and no straightforward E < R relationship could have been established.

of the English Simple Past and Present Perfect: For the Simple Past E and R coincide and are both situated before S, for the Present Perfect E is, of course, also before S, but R and S coincide. Thus Reichenbach has solved the problem of the similarity and difference of both tenses: the event itself is in the past in both cases, but the reference points are different.

Before we enter into a deeper discussion of the merits and problems of this system, we should reproduce some of his suggestive sketches, showing the location of the three points in each of the six major tenses of English:[9]

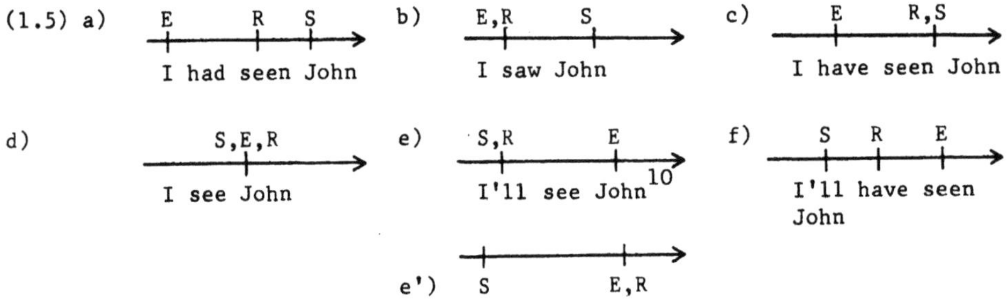

1.1.2 Aspect

For Reichenbach E, R and S are enough to handle not only the tenses, but also aspect. For the English progressive he suggests that "the event covers a certain stretch of time", (Reichenbach 1947:290) as in:

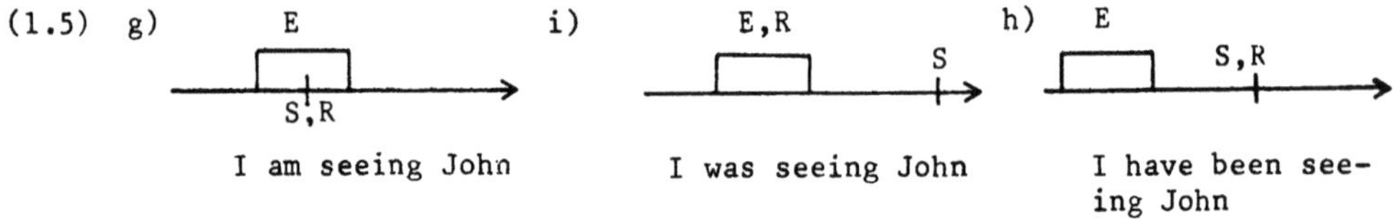

and for habitual readings of the progressive or the habitual aspect like the Turkish <u>muzari</u> he gives:

---

9 Actually, nine possible relationships of R, E, S establish nine tenses, that need not all be manifest in every language. We discuss only the basic ones and refer the reader to Reichenbach (1947:297) for more.

10 Reichenbach's analysis of the Simple Future is therefore not only a mirror of the Simple Past. The two analyses are necessary because of
i. Now, I'll see John (e.)
ii. I'll see John tomorrow (e'.).

(1.5) k)

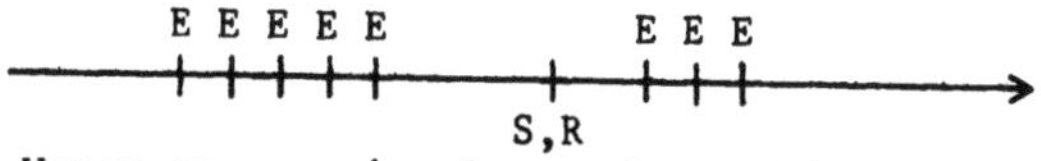

Women are wearing larger hats this year

görürü = I usually see

1.1.3 Criticism

There are some problems involved in this treatment, the sketches and the particular choice of examples.

1.1.3.1 Event notion

First, the choice of the verb *see* makes the analyses sketched rather special as (1.5.d) and (1.5.g) show. Activity verbs like *walk* would, for example, have a repetitive, habitual reading in the Simple Present, which couldn't be captured by d). For *I'm smoking* we would probably agree with a g) or k) reading, but *I smoke* can only be interpreted like k), never like d). Therefore the S,R,E-reading of the Simple Present appears to be a speciality of verbs like *see*, *hear* (a notoriously difficult group), the occasion of sports reports like *Jones passes the ball to Smith* and some uses of punctual events and achievements:

(1.6) a) Now I switch off the light
b) Now we reach the summit.

Reichenbach's analysis is, therefore, too broad, or - seen from the other side - not general enough.

1.1.3.2 Aspect

Second, there is ample evidence that duration is not the distinctive and only feature characterizing the English progressive. Concepts like "imperfectiveness" in contrast to "perfectiveness", "unbounded/bounded" or "bloblike" events have to be considered.[11]

1.1.3.3 Points or intervals

Third, at this early stage we already come up against the question of whether a reference time is a point or an interval. Reichenbach speaks of points and allows only the event and only in the case of the progressive to cover an

11 Cf. Comrie (1976:16).

interval. But in sketch (1.5) the rule of coincidence of R and E for a Simple Past seems to force R to cover an interval. Therefore it seems better to treat the three time entities as intervals, not as points. For adverbs like *yesterday* in contrast to adverbs like *at six o'clock* we need intervals in any case and also for the treatment of different event notions. Compare:

(1.7) a) * They ran a mile at six o'clock
b) They smiled at six o'clock.

It seems clear that we'll have to accept intervals as reference times.

1.1.3.4 The Present Perfect

Fourth, the treatment of the Present Perfect is deficient. Reichenbach gives another sketch for Present Perfect, that of "duration of the event ... up to the point of speech" (Reichenbach 1947:292). Though this is a common grammar book description of the 'continuative' Perfect it is simply wrong for Reichenbach's example:

(1.5) 1)

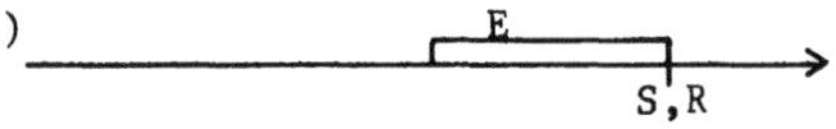

I have seen John

The event of seeing John does not last up to the present. Reichenbach is not able to capture experiential and resultative perfect just by the E S,R possibilities.

1.1.3.5 Semantic or pragmatic reference time?

Fifth, there is a certain vagueness about Reichenbach's use of his concept. We can well see what R is doing in the various analyses, but this does not define its status any too clearly. Is it a theoretical necessity (for the distinction of Simple Past and Present Perfect), or is it psychologically real (for the Past Perfect and on a text level)? And more immediately: Is it on a par with S and E as a semantic concept relevant for the truth-conditions for temporal expressions or is it a pragmatic concept not influencing the truth of a proposition?

Reichenbach's use of R seems to show that this third time position is in status not different from S or E. The text, however, suggests a pragmatic concept. Commenting on Keats's: *Much have I travelled in the realms of gold ...* Reichenbach says: "we notice that here obviously the past events are seen, not from a reference point situated in the past, but from a point of reference which coincides with the point of speech" (Reichenbach 1947:289), thus making the reference point into a sort of vantage point from which events are seen. From here it is not far to the statement that Reichenbach's R is a sort of speaker's perspective (Hinrichs 1981) and doesn't belong into semantic description at all (Partee 1984).

There are, however, some simple arguments that RT is relevant for the truth-conditions of a given sentence. Consider:

(1.8) a) When Thelma hurried past the bakery, she breathed a sigh of relief.
b) When Thelma had hurried past the bakery, she breathed a sigh of relief.

If RT didn't effect truth-conditions, both Past Perfect and Simple Past would only require that the respective events of breathing lie in the Past and then the meaning difference in the above sentences could not be captured: a) is true in a situation in which Thelma hurried past the bakery and breathed a sigh of relief simultaneously, while b) is true if she breathed a sigh afterwards (cf. Dinsmore 1982:221 f.).

Consider also:

(1.9) a) Tom borrowed $100 from me
b) Tom has borrowed $100 from me.

While a) could be uttered even if Tom has paid me back, b) appears to be a lie in that case (unless it is an experiential use).

Therefore it seems sensible to consider RT as a semantic concept. In which ways it has been treated semantically will be discussed later.

1.2 Reichenbach and adverbs

The concept of reference time is, however, not only relevant for the analysis of tense and aspect. For Reichenbach "it is the carrier of the time position" (Reichenbach 1947:294), and as such becomes important for any analysis of texts or discourse. R acquires this importance, because Reichenbach claims that "a time determination ... is referred not to the event, but to the reference time of the sentence" (Reichenbach 1947:294). Temporal adverbs therefore specify reference times not event times, and event times are qualified only indirectly via the coincidence of R and E. Compare:

(1.10) a) I met him yesterday
b) I had met him yesterday
c) I had met him before we left Freiburg. Then I didn't see him for five years.

In view of c) Reichenbach seems to be correct: Not the event of meeting counts, but its reference point. But as Reichenbach would only get the unusual reading for b) (the meeting took place before yesterday), this statement clearly has to be qualified, especially as it does not distinguish classes of time adverbs.

Some adverbs giving temporal information certainly do not have to do with the RT, but only with event time:

(1.11) We lived there *for two years*.

Reichenbach's statement therefore, though incorrect in its generality, turns out to be valuable as it provides a new parameter for the classification of adverbs.

One of the most important classifying distinctions is probably the fact that some adverbs are anchored to speech time (*tomorrow, now, yesterday*), others to reference time (*before, five minutes later, then*).[12] The former, the deictic adverbs, have been well investigated. But there are problems about the reference-time anchored adverbs:

(1.12) When we came home, John had already arrived. Ten minutes later we had a quarrel.

Ten minutes later clearly counts from the reference time, not the event time, i.e. from our coming home, not John's arrival. But there are more complex cases for which we need Couper-Kuhlen's idea of secondary reference:

(1.13) When we came home, John had already arrived. He had simply walked in. Ten minutes later he had finished the last of my beer and had started on my whisky. Well, we had a blazing row.

Investigation of adverbial function, classes and the adverbial role in the temporal constitution of texts is needed.[13]

1.3 Reichenbach and complex clauses

Equally interesting is Reichenbach's treatment of complex clauses. Reichenbach maintains that in a complex sentence with several clauses, the events described may occupy different positions in time, but that the "reference point should be the same for all clauses" (Reichenbach 1947:293), which

12 Cf. Bäuerle (1979), Harkness (1985), Schopf (1984), Smith (1978).
13 Harkness, this volume.

Reichenbach calls the *Principle of the Permanence of the Reference Point.*

Thus Reichenbach claims that in

(1.14) I had mailed the letter when John came in and told me the news

the three reference points coincide in the manner of the following sketch:

(1.14')

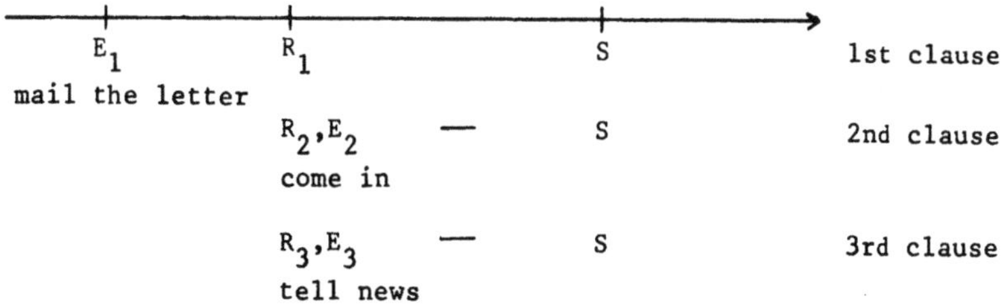

First off, this principle looks very convincing, especially as it automatically blocks:

(1.15) * I had mailed the letter when John has come ...

because here the reference points have changed. But it cannot explain sequence readings of Simple Past *when*-clauses:

(1.16) When John came in, he saw her at once

and furthermore, as Reichenbach notes himself, it creates a conflict in case of conjunctions like *after* and *before*, where we certainly can't have a permanent reference point. In those cases Reichenbach maintains that the *Positional Use of the Reference Point* is more general and is to be employed with *before* and *after*. According to Reichenbach, the conjunctions order not events but reference points: "In *he telephoned before he came* R_1 is said to be before R_2" (Reichenbach 1947:295). He admits that the two principles create a conflict in the following case:

(1.17) This is the man who drove the car

does not adhere to the permanence principle (which would require a Present Perfect in accordance with the use of *this*) because the addition of an adverb would violate the principle that adverbs specify reference points not events (which is part of the positional principle):

(1.18) This is the man who drove the car yesterday.

German adheres to the permanence principle and ignores the positional use, because the adverb specifies event time:

(1.19) Dies ist der Mann, der den Wagen gestern abend gefahren hat.

This possible clash of principles and the observation that natural language does not always obey the logical rules makes the point about (1.15) much less convincing than it looked at first glance. The remark about the erratic behaviour of language excuses almost any counterexample to the rules and also shows a somewhat prescriptive attitude. Concerning

(1.20) a) We'll dine when we have finished

Reichenbach implies that it would be more "logical" and "correct" to say:

(1.20) b) When we'll have finished ...

He concludes "that a mathematical language can be coordinated to actual language only in the sense of an approximation" (Reichenbach 1947:298).

As we already said, this pulls the rug very nicely from under the feet of any kind of criticism. But there are still a few points to mention.

It seems remarkable that Reichenbach treats all kinds of subordinate clauses on a par: relative clauses, *that*-clauses and temporal clauses with *when, before* and *after*. This and the fact that all clauses carry a full set of points, i.e. E, R and S, later led researchers to analyse them all in the same way (Smith 1978;1980;1981) or to maintain that for

(1.21) After they had eaten everything, they left

the main-clause RT and so the event of leaving provides the reference point for the event of eating (Nerbonne 1984:3). An *After*-clause is, however, an adverbial clause and should therefore functionally be treated like the adverb *yesterday*, so that it could never be the main-clause RT that provides the sub-clause RT for (1.21).

Reichenbach's formulation concerning the *before*-example leaves room for interpretation: "R_1 is said to be before R_2". In the best case we can assume that Reichenbach takes R_2 (the sub-clause RT) as some point of orientation from which the main-clause RT (R_1) is being located. Reichenbach's use of indices in the order of clause occurrence and his failure to mention subordination makes a generalization of this particular formulation rather difficult, however.

Clearly connected to this - the failure to distinguish main- and subordinate-clauses - is the next problem: a full set of points is allotted to all clauses.

This assumption predicts a meaning difference for all clauses with a Past Perfect and a Simple past. However, there are notable cases of similarity of meaning, cf. Quirk et al (1972:92), and

(1.22) a) After he had gotten home he called his lawyer
b) After he got home he called his lawyer

which is tending to lead to the disappearance of the Past Perfect in connection with *after*.

Two factors have to be considered here:

1. There might be a difference in independent and dependent Past Perfects, the latter being nothing but an embedded Past or Perfect,[14] therefore no longer involving the point of speech directly, while the former may need three time points for interpretation.

2. With temporal conjunctions it could well be the event of the subclause and not the RT which is relevant for the placement of the main-clause RT. Arguments for the point are measure phrases:

(1.23) He dropped the letter two seconds before he had reached the door.

Two seconds here clearly measures the length of time between the event of reaching the door (not its R) and the R (and event) of dropping the letter. It would not make sense to assume that *two seconds* is the time distance between the main-clause R and the sub-clause R, which in a Reichenbachian Past Perfect analysis would be later than the event - unless the analysis of the Past Perfect is to be rejected in favour of something involving the final point of the event as the natural R of a Past Perfect:

(1.23') a) false

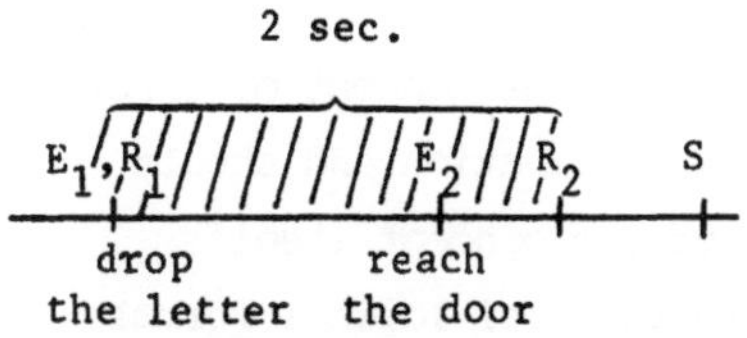

(1.23') b) correct

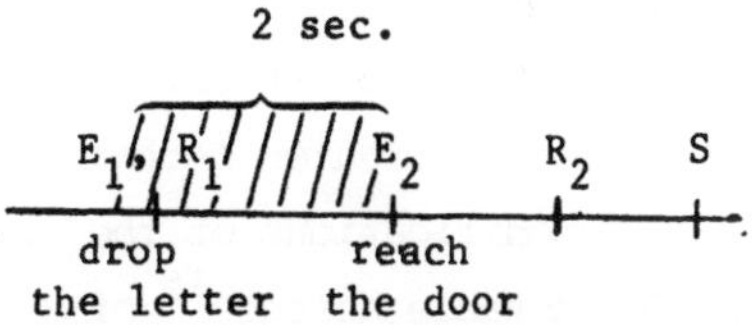

If one arrives at the conclusion that it is the event of the sub-clause that is relevant, not the RT, then it is only a short step to abandoning the sub-clause RT altogether and the assumption that the event is simply "given". This is more or less what Schopf (1984) does (cf. Schopf 1984:285). He postulates no separate RT for a temporal clause and speaks of clusters, i.e. one RT serves as orientation axis for more than one event. One argument for this view is that, on the text level, a reference-time anchored adverb in the next

14 Cf. Fabricius-Hansen (1984), Bäuerle (1979), Tichý (1980), Lyons (1977).

sentence can pick only one anchor, and this anchor is the main-clause RT, cf. (1.10.c) and (1.12) and

(1.24) He called me before he went to the airport. *And then* he must have mailed the letter.

This approach would get into difficulties, however, in those cases, where a difference in tense clearly signals a difference in meaning:

(1.25) a) He tore up the letter before he had read it
b) He tore up the letter before he read it

or

(1.26) a) When she got home she called a lawyer
b) When she had gotten home she called a lawyer

and the examples about Thelma's sigh of relief (1.8) a), b). Thus it looks as if for the treatment of temporal clauses we need at least a subsidiary RT, which being a sub-clause RT need not have the same information value on the text level as the current (i.e. main-clause) RT.

1.4 Summary and outlook

We have devoted so much space to the representation and discussion of Reichenbach's original system for three reasons:

1. Reference time as the "carrier of the time position" is obviously a very important and potentially valuable concept on more than one level, extending from an account of the tenses to an account of the temporal constitution of texts.
2. If applied to natural language in detailed examples the concept of reference time shows or creates certain problems that require explanation and should lead to modifications of the proposed analyses or of the concept itself.
3. The notable absence[15] of a clear (formal) definition, of logical formulas and the presence of sketches and rather vague verbal formulations make it necessary to give an idea of how the reference point works and what it was originally meant to capture before we progress to the question of what precisely it is, how it can be captured and defined, and how it can be fruitfully employed.

Parts II and III will investigate attempts to define the concept or else to abandon it in frustration, the modifications it has undergone (II), and the

15 The formulation is McCawley's.

recent comprehensive systems making use of it (III).

One fact has to be mentioned beforehand: Transformational Grammar hardly ever uses the reference point and it shows up only in logically and truth-conditionally oriented approaches (marginally in McCawley's variety of Generative Semantics), so that the pursuit of reference time is rather demanding for the untrained reader. We will try to keep the argument as simple as possible.

Part II Rank Weed or Fruitbearing Tree

2.1 Wild Growth

Before we turn to a more systematic discussion of the developments, let us note that the term is widely used in a rather intuitive way, so that it pays to check what the respective author really means when he/she uses the term. As examples let us discuss McCawley (1981) and Smith (1978).

2.1.1 McCawley

McCawley (1981) expressly promises answers to logical problems and so one turns to his chapter on tense with the hope of thorough enlightenment about what exactly the reference point logically is. But though the chapter certainly has its merits (cf. the considerations about branching time and the insufficiency of the usual tense operators) it is very confusing, if not misleading, with regard to reference time. On the other hand, some interesting points show up even in this formally deficient treatment, and we arrive at a rather clear idea of how the reference point cannot be treated. Readers not interested in blind alleys are invited to skip this section.

McCawley's tense logic is more modern than Reichenbach's in that it uses intervals and points,[16] and it proves the progress formal logic has made in discussing several possible ways of dealing with time.

The most naive way is to treat time, both points and intervals, as arguments of predicates and to build the dependence on time into ordinary predicates

---

16 Intervals are used for temporal concepts below sentence level, sentences are evaluated at points (McCawley 1981:345). Intervals are needed for adverbs like *yesterday*, but this cannot be the whole story on the analysis of adverbs: Compare i. They got married a year ago yesterday.
ii. They got married yesterday.

making an n+1-place predicate out of an n-place predicate: If Bald(x) means "x is bald", then Bald (x,t) means "x is bald at time t".

The same time dependence can be introduced via an operator R_t. R_t (Bald (x)) also means "x is bald at time t". From this paraphrase and equivalence we can deduce how the syntactic and semantic rule for R_t should look, even though such rules are missing in McCawley's treatment, making his R_t almost as vague and open to interpretation as Reichenbach's R. Using our common sense we get:

If ϕ is an (open) sentence, then $R_t(\phi)$ is an (open) sentence (and can be read: ϕ at t).[17]

$R_t(\phi)$ is true iff ϕ is true at t.

A third way to incorporate time is the use of temporal indices, which we need anyway for a unified theory of deixis. Then a proposition is not only evaluated at a possible world but also at a moment in time.

So far, everything is straightforward. Let us note two things, however. First, Bald (x,t) and $R_t(\phi)$ are "syntactic" solutions in the sense that they burden the formal syntax,[18] while the indexical approach is purely semantic: the indices (world, time and probably place) determine the model in which a formal expression is to be interpreted. Actually, syntax and semantics should always be developed side by side, so that they fit each other.

Second, $R_t(\phi)$ is true at those t where the event or state described by ϕ occurs or obtains, i.e. the operator singles out the event times of ϕ. In connection with a discussion of Reichenbach, we find however: "As a first approximation one could take 't is reference time for ϕ' to be equivalent to 't and ϕ appear in the combination $R_t(\phi)$ in the logical structure of the sentence'" (McCawley 1981:349). Even with the warning "as a first approximation" and the cautious modality of "one could take", this statement makes one wonder. Event time and reference time are different concepts in Reichenbach!

---

17 An open sentence is syntactically like a sentence, but contains variables. A sentence can be true or false, an open sentence can only be satisfied in a model. Bald (x) is an open sentence, Bald (j) with j for John a sentence.

18 Linguists tend to take the formula as the meaning. For the logician it is the syntactic side of the story and the more interesting things happen on the semantic side in model theory. The danger lies in the fact that some new syntactic device may not be easily integratable into a sensible formal semantics, so that it does not solve any (linguistic) problems. The formula can be taken as the "meaning" only if there is a canonical interpretation mechanism for its syntax.

Again, there is no remedy but to have a look at how R_t works. Let us consider some examples:[19]

(2.1) When I started teaching here, housing was fairly easy to find
(2.1') (ιt:R_t(I start teaching here)) R_t(housing be fairly easy to find)

As a sort of paraphrase we would get:[20]

(2.1'') For that t such that I start teaching here at t, housing be fairly easy to find at t.

To take another example:

(2.2) When John married Sue, he had already met Cynthia five years earlier
(2.2') (ιt:R_t(John marry Sue)) R_t(ιt': t - t'= 5 years) $R_{t'}$(John meet Cynthia).

Adding brackets in order to make the formula syntactically unambiguous,[21] we get:

(2.2'') (ιt:R_t(John marry Sue))R_t((ιt': t - t' = 5 years)$R_{t'}$(John meet Cynthia)).

As a rather clumsy paraphrase we have:

(2.2''') At that t such that John marry Sue at t, it holds that for that t' with t - t' = 5 years, John meet Cynthia at t'.

With the above first approximation of what reference time is, we get that t is reference time for (housing be easy to find) in (2.1) and that t is reference time for (John marry Sue), t' is reference time for (John meet Cynthia) and t is also reference time for ϕ = ((ιt': t - t' = 5 years) $R_{t'}$(John meet Cynthia)), however that is to be paraphrased. We see, of course, that this cannot be Reichenbach's concept any longer: t' would be event time for (John meet Cynthia) not reference time, only t could be reference time for that proposition.

The natural way out is to claim, that "reference time appears as a superstructure in which the remainder of the sentence is embedded" (McCawley 1981: 349). This, together with our bracketing in (2.2'') would make only t into

---

19 McCawley (1981:347), examples 11.3.13b and 11.3.14b.

20 We assume that in McCawley's syntax $\forall$, $\exists$ and ι (the iota-operator of definite descriptions, cf. Reichenbach 1947:§47) can bind the subscript t of the operator R_t and read the formula accordingly.

21 R_t has to operate over an open sentence, but (ιt': t - t' = 5 years) (that t' such that t - t' = 5 years) is a term! To put R_t in front of it makes no sense at all. We assume therefore, that R_t operates over the rest of the formula.

a reference time for (John meet Cynthia), but it would leave us flat in all those cases where we can't have an embedding, cf. (2.1). Apart from this, there is a worse problem, which McCawley himself mentions: Nothing really compels us to use the second R_t in (2.2'') because without it the formula's truth conditions would remain the same:

$R_t(R_{t'}(\phi))$ is true iff $R_{t'}(\phi)$ is true at t
iff (ϕ is true at t') is true at t
iff ϕ is true at t'.

The latter step is valid because " ϕ is true at t'" remains valid at any other t - if it is true.[22]

A rule which would compel the use of the second R_t in (2.2') and so arrive at an embedding would therefore, in this system, remain syntactic. It would not solve the problem of the Reichenbach-RT anyway, because only complex clauses involving embedded tenses[23] would be awarded a reference time.

Because of these problems, McCawley suggests a new system which unfortunately remains unsatisfactory, too. McCawley now incorporates times as superscripts, and these superscripts tie in with the indexical approach as they are supposed to give the times at which an expression is evaluated. As A^a translates into $R_a(A)$ if A is a timeless expression not involving any tense operator, we arrive at a reduction of "evaluation time" (the superscript) to "reference time" (in combination with R_a) to "event time", as we have seen when discussing the proper interpretation of R_t. McCawley does not notice this. He mentions something else: Given this new formulation "the reference point for ϕ is simply the point it is evaluated at" (McCawley 1981:350) and his new system of superscripts, only t' shows up as evaluation point in the ultimate formula for (2.2) and therefore only t' qualifies as reference point for (John meet Cynthia) - which is not at all surprising, considering the above "reduction" and the fact that t' is event time for that proposition.

These particular formalization attempts have, therefore, not clarified Reichenbach's concept and have certainly not captured it. On the contrary, they have created new concepts that are nevertheless all called reference points.

---

22 The problem here is that a semantic concept like truth at t is smuggled into the syntax via R_t without any reflection on the kind of model theoretic semantics we need for such an extension.

23 This is obviously what Tichý (1980) suggests for the treatment of tenses. He denies the existence of an unembedded Past Perfect.

There are one or two points to note, however. As Reichenbach pointed out some events can become reference times, i.e. a time already defined and identified on the time line can be used as reference time. This definiteness seems to be important and explains McCawley's (otherwise inexplicable) use of the iota-operator in ($\iota t{:}R_t$(John marry Sue)).[24] It cannot be the operator R_t itself which makes the event and the event time unique. This becomes clear if one considers that John could have married Sue, divorced her and married her again - something which becomes more plausible if one substitutes "J.R." for "John" and "Sue Ellen" for "Sue". Other kinds of events make the point more clearly:

(2.3) When I switched off the stove the water had been boiling for five minutes already.

There are many times at which (I switch off the stove) is true, but only one of them qualifies as the reference time here. So McCawley's use of the iota-operator is - as it stands - formally wrong, but intuitively correct: Reference time is normally a definite time, and this strongly points to a different approach, the indexical one.

Let us also note that a seductively easy way out of McCawley's dilemma concerning the definition of the refernce point, does not work either. Looking at (2.2'') one might be tempted to say: Everytime a time t is used to define another time t', it is a reference time. This idea would not work in the case of simple sentences involving only one time, at least not if one wants to stick to a Reichenbach analysis of tense, but it is nevertheless an often adopted use of the term reference time and leads to considerable confusion.

2.1.2 Smith (1978)

A case in point here is Smith (1978) who purports to use the term RT in Reichenbach's sense, sticking especially to the principle that adverbs set RTs. This leads to some rather astonishing analyses:

(2.4) Phyllis decorated the cake before midday.[25]

Smith (1978:49) comments: "In this sentence RT is Past, midday; ET precedes RT as indicated by *before*."

---

24 This operator may only be prefixed to an open sentence, if there is one and only one individual satisfying the sentence - i.e. we have a definite description.

25 Smith (1978:49), example (26).

This would lead to a sketch:

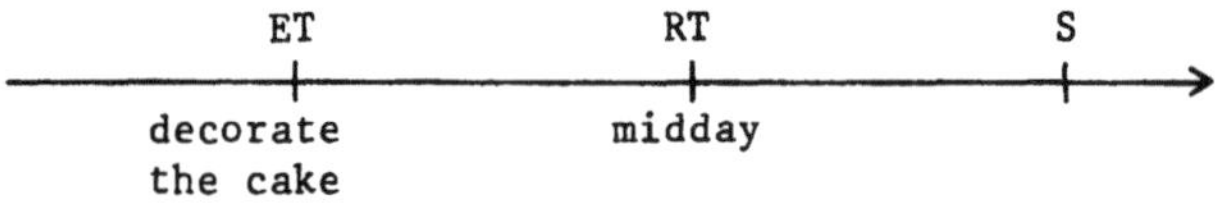

and so corresponds to a Reichenbach Past Perfect, not a Simple Past.

It is easy to see what has gone wrong: "Midday" is a time which serves for calculation (orientation, anchoring, definition) of another time, but it is not the whole adverbial phrase. The adverb is *before midday* and gives us a stretch of time in which the event took place - whether and in which way it also "gives" reference time will have to be discussed later.

Smith also speaks of an "anchor", but never abandons her use of the term reference time, which makes her otherwise very interesting papers (1978;1980; 1981) almost incompatible with other discussions of the reference point, in particular those analysing the tenses.

Let us keep in mind the fact that Reichenbach's term is frequently adopted in the literature, but not always in a Reichenbachian sense.

2.2 Rank Weed!

Reichenbach had to be taken into account, however, by any serious author writing on temporal systems, so that we are faced with a wealth of criticism.

2.2.1 Prior

Prior, whose tense logic is the foundation for most formal treatments of tense (cf. Montague 1974), used only two points, the event point and the point of speech. Complex tenses are analysed as embedded tenses and only the speech point serves as a constant. Event time is introduced via the existential quantifier in the definition of the past operator P (and the future operator F):

$$M \models_{t_o} P\,\phi \qquad \text{iff} \quad \exists t'(t' < t_o \text{ and } M \models_{t'} \phi)$$

where M is a model and t_o is the speech point. The formula can be read as: Past ϕ is true at t_o if and only if there is a t' earlier than speech time at which ϕ is true (always with respect to the model M).

There are well-known arguments against such an indefinite analysis, which requires only that some point in the past exist at which the event took place. These arguments will be presented later.

What interests us here is that Prior (1967) argued against Reichenbach's three points, maintaining on the one hand that the point of speech is conceptually not different from a reference point and just happens to be the first point of reference, and on the other hand that more than three points can be involved in the interpretation of a sentence like:

(2.5) I shall have been going to see John

for which he suggests something like

In a formalization of such problem examples Guenthner (1979) showed that it is true that an additional point (a point "from where" the reference point is reached) is needed to capture forms like (2.5). Reichenbach, however, still qualifies as a special case of such a complicated system, the case where the point "from where" is always the speech point. As the point "from where" and the "reference" point are distinct concepts in such a system the need for the reference point remains - distinct from speech point.

The whole problem vanishes, of course, if one does not require that the three Reichenbachian points be linearly ordered in all cases, a solution Nerbonne (1984) and Comrie (1985) suggest for the future perfect:

(2.6) I don't know whether he's left. He certainly will have left by tomorrow, however.

(2.6')

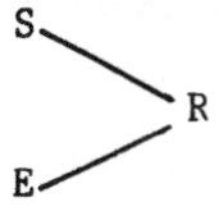

Thus no additional point is needed, we simply do not require that we can fully order the three points. If one adopts this view, one can maintain with Nerbonne that not more than three points are involved in the interpretation of temporal expressions - which does not put a limit on the number of points playing a role in the interpretation of complex sentences or texts.

On the other hand, this solution seems to carry us outside the normal conception of time, for the simple reason that if all of these three points are points of physical time they are elements of a linearly orderable set. There is one consideration in favour of this step out of time: the reference point in its more subjective interpretation as a vantage point might well be a point removed from the actual time line or be part of a counting line, but not a time, which might fit in with many considerations concerning the interplay of tense and modality.[26]

We will leave this question open and only recall that Guenthner's system captures Bull's and Prior's problem examples and still employs a reference point and a conceptually different additional point, not just another reference point as Prior seems to suggest.

The argument that the reference point is just another point of evaluation and therefore does not differ from speech point is not tenable either as Kamp (1971) has shown. Kamp's examples prove that no matter how deep the embedding, we still need the speech point as a separate anchor, it is not a matter of a simple shift:

(2.7) A child was born that shall be king.

2.2.2 Bäuerle, Fabricius-Hansen and the term "Betrachtzeit"

Yet the view that we can do without the reference point for exactly this reason is still widely held. Bäuerle (1979) maintains that for the simple tenses we do not need a reference point. For him the speech point suffices as a point of orientation and only the composite tenses need two points because they are cases of embedding the operator Perf which is itself not anchored to speech point.[27]

One of the major reasons to abandon the concept of reference time is that for the German tenses, at least in Bäuerle's analysis, a Present Perfect and a Simple Past are morphological variants of one another - on one reading. So the notation E = R is redundant, it does not serve for any useful distinction.

The case of the English tenses is, however, different: there is no reading of the Present Perfect where it could be called a morphological variant of the Simple Past. Therefore it might be wise to keep the concept of RT and the E,R — S and E — R,S analysis of these English tenses. This we should keep in mind when discussing other systems.

The most recent and most extensive treatment of the German temporal system, Fabricius-Hansen (1984), makes the same basic assumptions as Bäuerle. Both Fabricius-Hansen and Bäuerle abandon "reference time" in favour of Bäuerle's term "Betrachtzeit". The idea is, that Betrachtzeit is exactly that span of time that is being specified by an adverb like *yesterday*, which is a

26 Cf. Matthews forthcoming or, for another approach, the branching-time models.

27 Nerbonne (1984) follows Bäuerle in so far as he postulates only event time and reference time as necessary points for the perfect infinitive.

Betrachtzeit-adverbial in their terminology.

Because Reichenbach said that adverbs provide reference time, Betrachtzeit and reference time have been said to be the same thing (Hinrichs 1981:21). Bäuerle insisted that they are not. Given the different ways the tenses are analysed by Reichenbach and Bäuerle, and the roles reference time and Betrachtzeit play in these respective analyses, Bäuerle is certainly right. Bäuerle's "Zählzeit" or evaluation time, which he needs to account for Present Perfect and Past Perfect, also looks somewhat like reference time, at least when one compares the Past Perfect analyses. So we have to conclude that Reichenbach's R is a bit like Zählzeit (the time from which the event of a Past Perfect is located or the time from which an adverb like *later* counts) and a bit like Betrachtzeit because of the connection to adverbs.

Let us take another look at the interplay of reference time, event time, adverbs and Betrachtzeit.

The basic dilemma is that the event time of

(2.8) Hans played tennis yesterday

is normally not all of yesterday, so that either Reichenbach's E = R analysis for the Simple Past is wrong or *yesterday* cannot specify reference time precisely.

In Bäuerle's and Fabricius-Hansen's analysis *yesterday* gives Betrachtzeit and the simple tenses only have the task of specifying the relationship of speech time and Betrachtzeit. What we need therefore is the placement of the event, and this is introduced in Bäuerle via the morphologically not necessarily manifest frequency adverb "once". This gives the following paraphrase for (2.8):

(2.8') Hans played tennis yesterday at least once.

Thus (2.8) is true iff (Hans play tennis yesterday at least once) is true at that part of yesterday which is before speech time (i.e. all of yesterday) iff

$\exists t'$ ($t' \subset$ *yesterday'* and Hans play tennis is true at t').[28]

This allows Prior's existential quantifier to range only over the times of the past which are included in *yesterday'*. It seems to be a neat solution, but isn't really, as we shall see later.

---

28 *yesterday'* is the interpretation of *yesterday*, i.e. the day before the day including the speech point. Bäuerle's operator Prät (preteritum) works like this: Prät ϕ singles out the maximal past interval from the given Betrachtzeit as the relevant interval for the truth of the proposition ϕ.

Fabricius-Hansen, in turn, does not assume a frequency adverb for all sentences, thus she does not automatically get event time via the existential quantifier. She comes to the conclusion that the "final" Betrachtzeit is not the whole interval given by the adverb, but only the contextually relevant part[29] (we might know Hans's usual schedule) and that a loose usage of frame adverbs like *yesterday* might exist where the adverb refers to "a part of itself",[30] a rather original thought.

Fabricius-Hansen also uses the terms "preliminary" and "final" Betrachtzeit. Analysing

(2.9) Anna spielte heute um 12 Uhr Tennis
played today at 12 o'cl tennis

she says: "*Today* ... specifies a preliminary BZ from which the adverb *at 12 o'clock* picks a certain interval and assigns this to the remainder of the proposition as the final BZ".[31] Fabricius-Hansen does not carry this any further, but given that event times very often are shorter than the period specified by the frame adverb, this process of determining a preliminary BZ and then picking out a final BZ should always be employed, even for sentences with no explicit, but only a contextual, specification.

The final BZ comes up in another connection as well, and then is not called final BZ but "betrachtete Zeit" (the time looked at). The definition of this term is: "A time which has to be looked at with regard to the truth of a proposition at that time will be called 'betrachtete Zeit' ...[32]

We see that the final BZ of the above examples and this betrachtete Zeit are one and the same. As a given time which has to be checked for truth, i.e. against actual event time, it is either identical to event time or to Reichenbach's reference time.

In other words, Fabricius-Hansen's treatment without indefinite event time for all sentences leads to the introduction of a fourth concept, that of betrachtete Zeit, in addition to the adverb-given Betrachtzeit - and the dilemma about the relationship of the adverb specified BZ (or RT) and event time remains the same, more or less.

We are therefore thrown back to Bäuerle's solution and the question of why an "indefinite" event time, even when restricted to a certain interval, doesn't work. The arguments against an existential analysis for all readings are a modified version of the arguments against Prior's unrestricted use of the exis-

---

29 Fabricius-Hansen (1984:63).
30 ibid.(1984:64), my translation.
31 ibid.(1984:40), my translation.
32 ibid.(1984:39 f.), my translation.

tential quantifier. I will discuss three of them:

1. The negation argument.

This is due to Partee (1973) to my knowledge but can also be found in Bäuerle (1979), Hinrichs (1981) and Nerbonne (1982).

(2.10) I turned off the stove

if analysed as

(2.10') $\exists t\ ((t < t_o) \wedge (\text{I turn off the stove at } t))$

could be negated with wide or narrow scope for the negation, as

(2.11) a) $\neg \exists t((t < t_o) \wedge (\text{I turn off the stove at } t))$

or as

(2.11) b) $\exists t \neg ((t < t_o) \wedge (\text{I turn off the stove at } t))$

neither of which would capture the meaning of

(2.12) I didn't turn off the stove

because (2.11a) means that there never was a time at which I turned off a stove and (2.11b) is true if there is at least one time in the past at which I didn't turn off a or the stove, which is certainly true of all times before my birth and a lot of other times - therefore the sentence is always true.[33]

Partee's argument shows that the sentence talks about one particular occasion of stove operation - the conversationally relevant one. The same (with one reservation) applies to a sentence like

(2.13) I closed the door at least once yesterday

(2.13'') $\exists t'((t' \subseteq \mathit{yesterday}') \wedge (t' < t_o) \wedge (\text{I close the door at } t'))$

can be negated in the same two ways as (2.10), as

(2.14) a) $\neg \exists t'\ ((t' \subseteq \mathit{yesterday}') \wedge (t' < t_o) \wedge (\text{I close the door at } t'))$

which would be true if on every occasion yesterday I left the door open (and leads to our reservation, because it is a possible reading of

(2.15) I didn't close the door yesterday)

or as

(2.14) b) $\exists t^* \neg ((t^* \subseteq \mathit{yesterday}') \wedge (t^* < t_o) \wedge (\text{I close the door at } t^*))$.

The latter is true in the following situations:

---

33 I omit as irrelevant the possibility of negating $(t < t_o)$.

there exists a time t* which is a time of yesterday and I did not close the door at that time (which makes the proposition almost invariably true) or there exists some t* which is not a time of yesterday at which I closed the door, which is a reading for (2.15) with heavy stress on yesterday, but does not cover the normal reading of (2.15), that of the relevant occasion at which I didn't close the door. Neither does any of the other possibilities.

2. The situational dependence - a variant of the negation argument.

Bäuerle rightly points out that

(2.16) Ede hat eine Pizza gegessen

can be true today and a lie tomorrow, depending on the situation of utterance. If the analysis of the sentence involved an unrestricted existential quantifier and the sentence is found true today, then it will remain so for the rest of all time.

From this Bäuerle deduces that Betrachtzeit is needed to specify the relevant time and that Betrachtzeit should be a definite time. Valuable as this idea might be (cf. McCawley's use of the iota-operator) it does not solve the problem completely, if one allows the existential quantifier to creep in elsewhere:

(2.17) Ede hat am 1.2. eine Pizza gegessen

might be true and it stays true for "all time" if one considers only one reading of that sentence, i.e. the indefinite reading, where the whole of that specific day is under consideration. But on the other hand it might be false if he ate a pizza at lunch time on that day, but what we are talking about is his dinner.

Therefore we come to the conclusion that the existential reading is not the only one and that for the other reading Bäuerle's Betrachtzeit, though definite , is still too vague.

3. The sequence argument

One of the arguments against Prior takes us beyond a simple one-event sentence and ties in with what Reichenbach had to say about the context dependence and time-carrier-function of his reference times.[34]

(2.18) John went to Mary's place at eight o'clock, ate a bite and went straight to bed.

---

34 Cf. Nerbonne (1982), and also Hinrichs (1981:32).

Undoubtedly, we have a sequence of events here. How exactly one arrives at this sequence is still open to debate,[35] but it seems clear that we have to take into account clause order, semantic information about tense and event notion, and pragmatic facts. What is obvious, however, is that a semantic analysis à la Prior would leave all the work to pragmatics, since we would get "there is a t, at which John went to Mary's, there is a t', at which John ate a bite, and there is a t'' at which he went to bed", and nothing whatever constrains t, t' and t'' to any particular order. Pragmatically too, we get little help:

(2.18') He ate a bite (in a diner), went to Mary's, and went to bed.
(2.18'') He went to Mary's, went to bed, and ate a bite (which Mary served on a tray in the bedroom).

are just as likely sequences. That it might be therefore just clause-order that produces the correct sequence is highly improbable in view of:

(2.19) The bomb exploded. The ceiling crashed, the window shattered, plaster flew about ...

where we do not have any sequence at all. The existential analysis simply cannot cope with the order of events.

The same applies if we were to add *yesterday* to (2.18):

(2.18) b) Yesterday, John went to Mary's at eight o'clock, ate a bite and went straight to bed.

Though all relevant times are restricted to times of yesterday, there is no way to establish their order if we have a Bäuerle analysis.

This proves beyond doubt that we need some point as the carrier of the time position and that the positioning of times has to be more exact than it is possible to achieve with the help of speech time, Betrachtzeit, and an existentially introduced event time. So Bäuerle is seeing things too narrowly, when he claims to be able to do without a reference point for the simple tenses, since tenses occur in sentences and sentences occur in texts.

The discussion has shown two things so far:

1. Though the function of temporal adverbs such as *yesterday* is connected with the function of reference time, not event time, and so far Reichenbach was right, nevertheless adverbs and reference time are better kept apart. Adverbs provide a temporal setting which serves to locate reference time, but they do not provide reference time directly, and in this respect Reichenbach was

35 Cf. Schopf (1984) and Couper-Kuhlen, this volume.

inexact.[36]

2. There are two readings of

(2.20) Hans had a pizza yesterday,

an indefinite reading, probably involving the existential quantifier, and a definite reading, presupposing a definite occasion included in the interval specified by *yesterday* and therefore a definite reference time.

If there is no frame at all, i.e. not even a vague specification of the time, it is interesting to note, that an English Simple Past sentence tends to have the definite reading - probably because of the contrast to the Present Perfect, and here German and English seem indeed to differ:

(2.21) a) Hans had a pizza
(2.21) b) Hans has had (a) pizza.

2.3 Fruits

From this observation it is only a short step to introduce reference time into our models as an index on a par with the indices of speech time and event time. En route to this approach were Åquist and Guenthner. But Åquist (1976) needs two models, one which remains anchored to speech time, and one which we obtain through a tense operator. In the latter model reference time serves as an index, but it replaces the t_O (S or SP) of the original model. This shows that we are still working with the idea of shifting points (the operator also turns *now* into *then*) and the point of speech is indeed a kind of first reference point. A more recent formal model, Saurer (1984), fixes "frame time" (t_F) as an evaluation index, but like Bäuerle (1979) keeps the indefinite event time analysis.

To have three points from the beginning, however, makes it possible to check the truth of a sentence against a model in which these three times are already fixed, i.e. definite.[37] Assuming for the time being that Simple Past

36 Cf. Harkness, this volume, for the term "temporal setting" and another discussion of the adverbs/frame/reference time -problem.

37 There are some problems concerning the treatment of the English Present, which does not seem to be indexical, as there is no way of determining future or present reference without an adverb. Obviously, we cannot simply assume a definite occasion, we need to specify it. Other problems involve habitual and iterative readings and concepts like *work in a factory*. Such event types are normally not treated (cf. Nerbonne, Hinrichs), but Dowty (1982) makes some suggestions concerning habitual readings.

and Past Perfect are sentence operators,[38] one would get truth conditions in the same vein as Nerbonne (1982;1984):

(2.22) $\mathfrak{A}_{s,e,r} \models$ Past (p) iff (e = r and e < s and $\mathfrak{A}_{s,e,r} \models$ p)

(2.23) $\mathfrak{A}_{s,e,r} \models$ Past Perfect (p) iff (e < r < s and $\mathfrak{A}_{s,e,r} \models$ p)

with the proviso that for atomic p (i.e. for a proposition without any temporal expression) we do not need reference time for evaluation, so that $\mathfrak{A}_{s,e,r} \models p$ iff p holds at e.

There is some debate on whether Past Perfect should have a definite or an indefinite reading, or both.[39]

But leaving this problem aside there remains the task of incorporating adverbs into this framework without creating the old problems. One way is definitely not possible: we cannot let the adverb A substitute the index for reference time. Given a classification of adverbs into duratives, frame or Betrachzeit-adverbials etc. the case of Betrachtzeit-adverbials is such that a definition like

(2.24) $\mathfrak{A}_{s,e,r} \models$ A (p) iff $\mathfrak{A}_{s,e,A'} \models$ p

(in which A is assumed to be a sentence operator and A' is the time specified by A in our model)
creates all the old problems about event time covering not all of the interval specified by the adverb.

If we take into consideration, however, that, for this particular class of adverbs, our reference time, though definite, may be unspecified, we can take the frame (or Betrachtzeit) adverbial as a specifier or modifier and formulate the truth conditions for this class of adverbs like this:

(2.25) $\mathfrak{A}_{s,e,r} \models$ A (p) iff (r $\subseteq$ A' and $\mathfrak{A}_{s,e,r} \models$ p).

Note that

(2.26) Hans played tennis yesterday

---

38 If they are verb-phrase operators, and there are good reasons to believe that they are, we need only slight modifications to give similar satisfaction conditions.

39 Schopf (1984:332) opts for an indefinite Past Perfect and so does Nerbonne (1982) for examples like: *I talked to Fran. She had seen Bob.* But because of the event sequences in flashbacks told in the Past Perfect, Nerbonne (1984) and Hinrichs (1981) argue for a definite Past Perfect . Hinrichs introduces a second referene time to achieve this, the definite time against which the Past Perfect event has to be checked for truth.

is now true if the definite occasion talked about is in the past, is a part (or all) of yesterday and coincides with the time of Hans's playing tennis - which is what we wanted to achieve.[40]

On the whole the three index approach seems to be a very neat solution and it only requires a slight alteration of Reichenbach's original statement that adverbs provide reference time - not event time. This class of adverbs modifies reference time in the sense that they help to locate it on the time axis.[41]

The only problem now is the existence of indefinite readings. These we could treat by the insertion of 'at least once' (as Nerbonne (1984) suggests) or by abandoning the unified treatment and allow definite and indefinite tenses with analyses of their own or by burdening the rules of context shift. How the problem is best handled must remain open here.

Part III New Branches

So far our investigation has shown that

a) Reichenbach's original concept and tense analysis can be captured formally in such a way as to overcome the adverb - RT -ET problem,

b) that it might be wise not to abandon the reference point in favour of other concepts for the following reasons: 1. the neat way it can deal with the English tenses, 2. the definite readings of most tensed expressions, 3. the problematic interplay of tenses and adverbs and 4. the text-linguistic time-carrier function of the reference point.

As we have seen in 2.3 the indexical approach can nicely handle 1., 2. and 3.[42] But it keeps us on the level of truth-conditions. With the indexed model against which the truth of a sentence is checked, it assumes <u>that</u> the reference point is given, but it does not tell us <u>how</u> we arrive at this crucial point or how we progress from model to model.

---

40 Saurer (1984:47 f.) discusses the merits of the "frame" or modifying approach $t_F \subset t_{adv}$, in his terminology. He analyses *John smoked yesterday* and concludes that it is undesirable that the sentence should turn out to be false in the following situation:

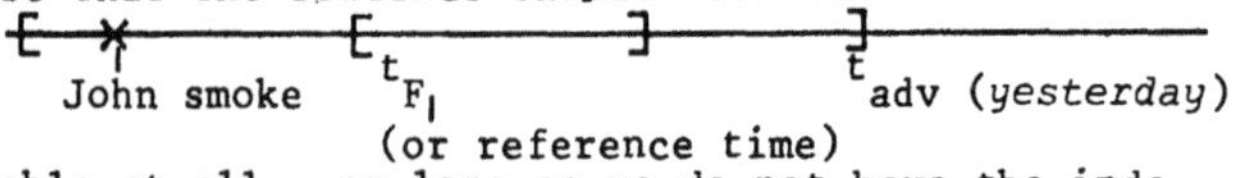

We do not find this undesirable at all - as long as we do not have the indefinite reading, but the one involving a particular occasion.

41 If we accept Hinrichs' second reference time (cf. 39) there is no exception to this rule. The two readings of
i. Hans had come on Tuesday
are explained because there are two reference times which can be modified.

42 Th: rule for Present Perfect is obvious:
$\alpha_{s,e,r} \models$ PresPerf p iff ($e < s$ and $r = s$ and $\alpha_{s,e,r} \models p$).

As long as we are only after the truth-conditions of a sentence, this is enough, as the often quoted similarity with personal pronouns proves (Nerbonne 1984, Partee 1973). But if we want to give the reference point its due as the carrier of the time position, we must investigate how it is being fixed and how we progress from one to the next. This takes us from the level of sentence semantics into that of discourse and therefore into a fairly new field of study. It also takes us "below" the level of the sentence because tense, adverb, and event notion singularly and combinedly all contribute to the setting and progression of reference points.

3.1 A simple theory

To give a rough idea of what is involved we will sketch a rather general approach and point out its merits and shortcomings.

To his indexical semantics Nerbonne (1982;1984) adds a pragmatic principle designed to take care of the progression of reference times and the sequence of events we can establish from what he calls "temporally connected discourse".

In its simplest form this pragmatic principle, called Reichenbach Pragmatics, appears as:

> RP: For S_1, S_2, ... S_n a sequence of sentences uttered in a temporally connected discourse,
> RT of S_i does not extend beyond RT of S_{i+1} $(1 \leq i \leq n-1)$.[43]

The obvious merit of this statement is that we can keep apart the semantic level, in particular the contributions of individual items to the meaning of a sentence, and the pragmatic factors like context or occurrence in a narrative or descriptive text. Note also that, following Reichenbach closely, the principle applies to reference points, not events. For

(3.1) Al went to New York. Bo had found him a room. He went directly to it.

the event of finding the room can be earlier than Al's arrival, but the reference point of S_2 is not earlier than that of S_1, it is probably the same, depending on the interpretation of S_3.

---

43 For formal definitions of "precede" and "does not extend beyond" see Nerbonne (1984:7 f.).

3.2 Event Notion

Very clearly, however, the principle is not specific enough. It only tells us that we never have regression of reference time, but it does not tell us when we have genuine progression and when we have simultaneity or overlap.

To capture the obvious difference in the temporal interpretation of the following two sequences:

(3.2) Al went to New York. The others went there, too.
(3.3) Al went to New York. The others were there, too.

Nerbonne (1984) adds an implicature for atelics:

RP' ... for atelic S_{i+1} it is implicated that RT of S_{i+1} does not extend beyond RT of S_i ($1 \leq i \leq n-1$).

Now the two parts of the principle add up to the implicature of simultaneity of reference time in the case of atelic predicates. This seems to be too strong, as it rules out an overlap situation. It can easily be amended to the requirement that RT of S_i and RT of S_{i+1} be non-disjoint for atelics.

But technicalities aside, the interesting fact is that we have to take account of event notion.

Schopf (1984) and Couper-Kuhlen (this volume) go into this problem, showing that it is not mere telicity that forces a sequence reading and atelicity that leads to simultaneity readings. Their line of investigation arrives at a more detailed analysis of the role of event notion in connection with the setting of reference times and of the problems of where to place reference time with respect to the phases of an event or with respect to an event given in the progressive mode.

(3.9) At midnight John was asleep

clearly shows that an ET = RT analysis of the Simple Past à la Reichenbach conflicts with the $RT \subseteq A$ analysis of Betrachtzeit-adverbials we have opted for, unless we consider the fact that we are dealing with a state and modify the analysis accordingly. We will have to reinterpret Reichenbach's simultaneity of event and reference time as inclusion of reference time in event time or event time in reference time, depending on event notion.

As *at midnight* in (3.9) not only gives a vague frame for the reference time, but identifies it exactly (with additional information about the date) we see that the reference time for states is included in the event time. For a quantified process like *smile* the reference time seems to include the event, making it both bloblike and simultaneous.

These facts seem to have been noticed almost simultaneously by Schopf and Hinrichs.[44] Hinrichs (1981) distinguishes a) states, activities and progressives, and b) accomplishments and achievements, grouping the first three together because they include their reference times and the latter two because they are included in their reference intervals. Partee (1984) simplifies Hinrichs to a distinction between states and events mirroring the above two groups. She also analyses progressives like states, i.e. as event notions including their reference times. Schopf (1984) distinguishes more event notions, which is necessary in view of a predicate like *smile*, which is not a simple process (or activity) like *walk* because it can be paraphrased as *give a smile*, which *walk* can not, and has an initial and a final point, which *walk* does not. But neither is *smile* an achievement like *reach the summit* nor an accomplishment like *run a mile* by the classic test (If you were smiling you have smiled, but if you were running a mile and were interrupted you have not run a mile). *Smile* is called a quantified process in Schopf (1984) and is distinct from other event notions.

For states, progressives and accomplishments Schopf and Hinrichs agree, though Schopf is careful to point out that states, especially, can be reinterpreted as other event notions.[45]

States are visualized as

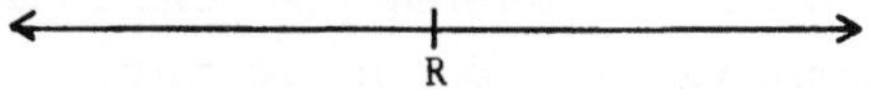

in Schopf (1984:99)

and accomplishments like *John drew a circle* as

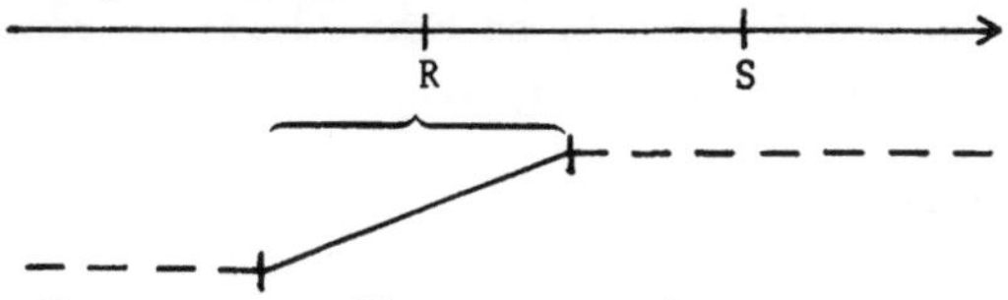

ibid., 240.

while the corresponding progressive *John was drawing a circle* is sketched differently

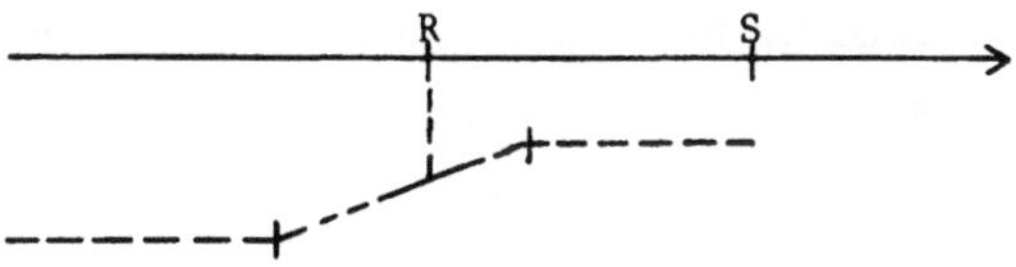

ibid., 240.

---

44 Schopf (1984), developed out of a manuscript available since 1981.

45 Cf.Example 3.5 and Schopf (1984:99). The question therefore is: At what level do we mark something as a state, at verb level, at sentence level?

Clear as these sketches may be (including or excluding initial and final points) there is one problem or advantage: the use of the brackets drawing together the reference time to a point. This gives RT a certain (intended) ambiguity: it is an interval comprising the event, but it can be conceived of as a point for those cases where events are viewed as blob-like or are simply being named.[46]

Now the analysis of *smile* yields:

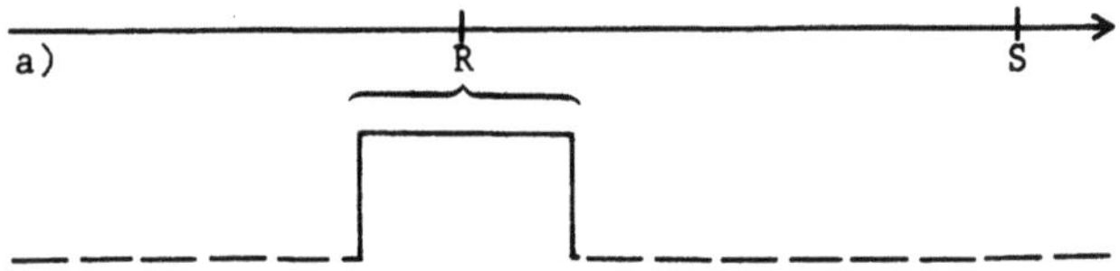

for *he smiled*, and

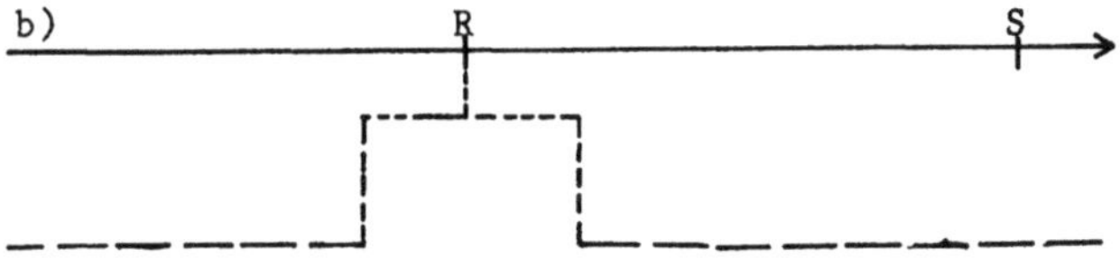

for *he was smiling*,

also placing the Simple Past event within the reference time as for accomplishments. So far, Schopf's more detailed analysis corroborates Partee or Hinrichs: The important distinction of event notions with regard to reference time is whether the reference time includes the event or the event includes the reference time. Other distinctions such as combinability with duratives, do not count here, because duratives do not modify reference but event time. There is one problem, however: If the inclusion behaviour is the only distinction of event notions with regard to reference time, how can it happen that *smile* can be blob-like, but *run a mile* cannot, so that (3.11) is acceptable only in the interpretation of an inceptive predication.

(3.10) She smiled at six o'clock
(3.11) * She ran a mile at six o'clock.

Is only our pragmatic knowledge about the normal length of event times or is some structural difference in the event notions responsible for this difference? And there are some more problems. Consider predicates like *switch on the light* or *look away*, which both assert a punctual change, presuppose an initial state and imply (if not assert) a resulting state. They differ in the interpretation of their progressives. While

(3.12) John was switching on the light

---

46 Schopf (1984:266). In connection with the present tense he identifies a subtype called "nennendes Präsens".

can have an interpretation where John has to work more than one switch to achieve illumination or has an interpretation involving conation (observe that (3.12) seems still to be true if John was on the way to the switch or had his hand on the switch but dropped dead before he could operate it),

(3.13) John was looking away

normally points to a reference time included in the resultant state. So here the simple distinction of what includes what is not enough. The reference time of the progressive does not simply sit somewhere within the event, it is bound to a certain phase of the event.[47]

For the simple form the sketch for both event notions

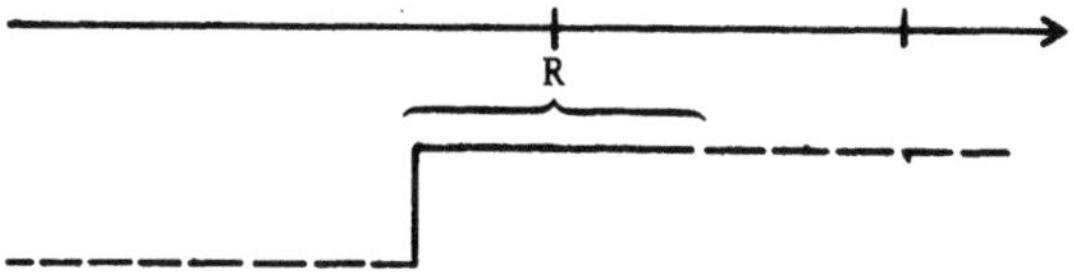

Schopf (1984:242)

has to be reconsidered. In view of the fact that point adverbs pin down reference time exactly and can only pick the moment of change, not a moment of the resultant phase, the RT should ultimately single out only the mutation phase of *switch on the light*:

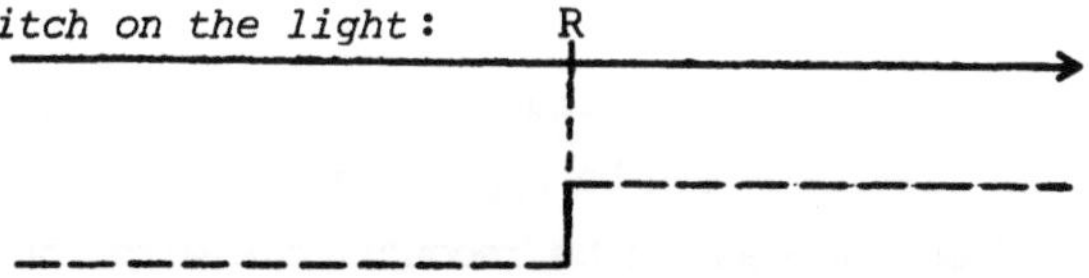

The fact that we mean the resultant state when we say

(3.14) John switched on the light for half an hour

need not bother us, because *for X* is a durative adverb. On the other hand, it might well be the case that we mean to comprise the change plus the resultant state when we utter

(3.15) John switched on the light yesterday

in a situation where it is important to prove that the lamp was in working order yesterday. For such a meaning Schopf (1984:242) would be correct.

A combination with a reference-time-dependent adverb, however, again ties the adverb to the moment of change:

(3.16) John switched on the light. 10 seconds later - exactly the time a careful man might need to take aim - a bullet smashed right between his eyes.

47 Cf. Saurer (1984), who requires for achievements that they be evaluated at the final point, or, better, the moment of achievement.

For *look away* the adverb and so the reference time is tied to the resultant state.

Some event notions therefore seem to have prominent points that are tied to reference time - no matter what other phases are involved in the event. These points seem to be relevant especially for the temporal constitution of texts.[48]

3.3 Discourse

Though Schopf (1984), which is based on these distinctions, has some interesting observations concerning the interplay of event notion, reference time and narrative progression, they are too unsystematic to make predictions on what triggers or blocks RT progression. Partee (1984) purports to give a general and simple rule for the setting of RTs, letting tense do the work, more particularly the use of a Simple Past in a narrative text.

The principle can be reduced to this: Given the narrative mode and a sequence of past tense sentences, each sentence with an event predicate places that event within the then given (current) reference time and automatically shifts reference time forward to just after the event. For a sentence with a state or progressive predication the state includes the then current reference time and no shift takes place.[49]

The very simplicity of this makes it attractive and easily assailable. Obviously the same kind of criticism can be applied as to Nerbonne's equally simple principle, and it does not seem to be reconcilable with the example:

(3.17) a) The bomb exploded. b) The window burst into a thousand pieces, the ceiling cracked, furniture flew around ...

where we cannot analyse the b) events as subevents of a) (which possibility Partee calls a break of the narrative mode) and there is simultaneity or at least no given order of events.

If we turn round and abandon the progression principle altogether, requiring only that a past tense is always to be anchored within a given reference time as Fabricius-Hansen does,[50] then we can easily explain (3.17) or (2.19) or a passage like:

---

48 Cf. Couper-Kuhlen, this volume.

49 Partee (1984), especially p. 253 and 255.

50 Fabricius-Hansen (1984) introduces speech time, Betrachtzeit and Co-time as indices and requires that Simple Past is Betrachtzeit or Co-time anchored. Co-times are contextually specified times.

(3.18) It snowed on the first of April. Traffic stalled and children got out their sledges.

But we cannot explain the sequence reading of example (2.18), (3.6) etc.[51]

It looks as if one has to make use of both principles and has come back to the question of which signals trigger which principle.

And even without this difficulty there are intrinsic problems which once again show the value but simultaneously the elusiveness of the reference point when used in complex systems like Partee's.

3.4 Bitter Lemons

We have seen that the reference point is necessary and valuable for

a) the analysis of tenses,

b) the understanding of the function of adverbs,

c) the distinction of event notions, and

d) the keeping of the time position in discourse.

The two formal analyses that try to take into account all of the above areas are Nerbonne (1984) and Partee (1984), which to some degree are variants of one another, though Partee does not distinguish the semantic and pragmatic level, in fact claims that it is time to develop dynamic semantics in the manner of Kamp (1981). Dynamic semantics incorporates semantic and pragmatic facts and functions into so called discourse representations. The embedding conditions that specify when a (temporal) discourse representation is embeddable into a model are similar to what Nerbonne gives as satisfaction or truth conditions as they "simply require that there be some events (or intervals) r_1, r_2, r_3 which satisfy the specified ordering conditions ..." (Partee 1984: 255).[52] Thus it suffices to analyse Partee (1984) which is more easily available for the interested reader. Partee (1984) in turn summarizes and develops Hinrichs (1981) so that we may consult Hinrichs on those points where Partee leaves us in the dark.

---

51 Fabricius-Hansen cannot solve the problem of the "Präteritum als Erzähltempus", which she discusses on p. 99. Her rule for preteritum p. 199, especially b.ii, which is designed to take care of the sequence readings, does not lead anywhere as a careful application of that rule and the rule for temporally neutral propositions on p. 202 to the example on p. 91 shows. The trick that pushes the BZ-index into the place of the evaluation time (normally occupied by speech time) simply doesn't work.

52 "Event" is here used in the broad sense and as a primitive. Intervals and points are defined via the intersection of events. Cf. Kamp (1979) for the formal theory.

3.4.1 The role of the past tense

Let us first look at an isolated past tense sentence like

(3.19) Sam left

or the famous opening of a novel:

(3.20) He lay flat on the brown, pine-needle floor of the forest, his chin on his folded arms, and high over head the wind blew in the tops of the pine trees.
(Hemingway)

Smith (1978:48) claims that a past sentence like (3.19) needs an anchor and is semantically incomplete because it is neither adverbially nor contextually anchored. Whether one agrees with this claim largely depends on one's concept of semantics and the parallel case of personal pronouns.[53]

On the other hand, it is quite clear that Smith is referring to what has sometimes been called the anaphoric past, which uses as reference point a contextually given point. The incompleteness here would then stem from the fact that there is no such point in (3.19) and (3.20). We have seen that Nerbonne blithely ignores this lack of a reference point. He interprets (3.19) by assuming one. There certainly is a vagueness about the interpretation of the sentence, because there is no other restriction given in the satisfaction condition except the one about a past reference point. Therefore (3.19) would be satisfied in infinitely many models and the addition of an adverb would narrow down the number of possible models, reducing vagueness in the sense of Kamp (1975). The assumption of a reference point is clearly legitimate. It seems to be exactly what one does, when confronted with (3.20), and it is nothing new. Lewis (1979) suggests the same procedure in the case of presuppositions. Partee (1984) also takes the same path. At the opening of a discourse a reference time r_o is assumed and the first event is then located with respect to this reference time. So a past tense is (in Partee 1984, not in Partee 1973) always anaphoric, does not carry its own RT and has to be anchored to a given though possibly created RT. On the other hand, a past tense sets the next RT, which makes it ambivalent with regard to reference time and its basic anteriority or posteriority direction.

---

53 I disagree, applying the same kind of argument that is usually applied to the semantics of personal pronouns. It is enough to know, that "I" means the speaker, it is not necessary to fill in the situation and identify the speaker.
Apart from this, I ask myself if Smith would call formulas involving the existential quantifier semantically incomplete, if it is not possible to pin down the individual that satisfies them.

Checking Hinrichs, we find exactly the same. For a past tense occurring in a main clause, we introduce the event into our discourse structure noting that it is included in the current RT in the case of accomplishments and achievements and including the current RT in case of activities, states and progressives, and then - in the case of achievements and accomplishments - we introduce a new interval after this event which is to be the next reference time.

The problem is that we do not get a basic meaning for the past, but a very specific one bound to the conditions of occurrence in a main-clause, to narrative mode and to event notion: only past events introduce RTs, past states do not. So what is the basic role of the past tense?

That this really is a problem can be seen when Hinrichs gives rules for temporal clauses and adverbs: He has to treat Simple Past in *after*, *before* and *when*-clauses separately and differently and then has to duplicate the whole routine for the Past Perfect. We neither get a basic meaning of *after*, nor a basic function of Simple Past.

Considering the examples collected by Schopf (1984) and Couper-Kuhlen (this volume), it is at least doubtful that it is the Past (coupled to certain conditions) that is responsible for the RT progression.

3.4.2 Adverbs and RT progression

The other area where we can expect trouble is that area where Partee's tense triggered progression RT meets with an adverb speciifed RT.

Discourse parts like

(3.21) He arrived at ten o'clock completely exhausted. The next morning he woke up rested and refreshed.

clearly show that no RT "just after" coming home can serve for the localization of waking up.

Partee makes a couple of remarks concerning this problem, i.e. the function of an adverb in general and especially in the above case. Her remarks are unfortunately somewhat vague so that one is not quite sure whether in her system the adverb modifies a given RT or actually introduces the RT (as in Reichenbach). She maintains that temporal clauses function like adverbs (which is certainly correct) so that what she says about subordinate clauses should carry over to adverbs and vice versa. Compare:

"... the adverb itself whether phrasal or clausal, provides a <u>descriptive characterization</u> of the new reference time" (Partee 1984:257) and "<u>tells</u>

where the new reference time is located relative to some event ..." (the case of an adverbial clause, Partee (1984:263)), which point to the modifying frame approach in the manner of Nerbonne, and

"... the subordinate clause ... introduces the reference time" (Partee 1984:263) or

"... the primary function of a *when*-clause ... is to provide a new reference time for the associated main-clause" (Partee 1984:261), both of which see the adverbial as the source of the new reference time.

As for the problem example (3.21) we find:

"If a sentence begins with a past adverb ... that adverb is processed before the main-clause and serves to update the current reference time. Thus it's optional that there be an adverb, but if there is one it overrides the prior context " (Partee 1984:257).

Clear as that passage might seem, it is not at all obvious what is meant by "update" and "override". "Update" also occurs on p. 258: "The *when*-clause triggers the updating of the reference time to a new value, let's say r_2 and the content of the *when*-clause puts conditions on r_2." Does this imply, that updating is nothing but the triggering of the RT progression?

Running through Partee's

(3.22) Mary turned the corner. When John saw her, she crossed the street.[54]
S_1, e_1 S_2 , e_2 S_3 , e_3

we find that e_1 is anchored to r_o (created) and triggers the setting of r_1 as expected. But then something curious happens: the *when*-clause introduces its own RT, called r_2, the *when*-clause event is anchored to this RT: $e_2 \subseteq r_2$ and r_1 is only needed to place r_2 after r_1. This does not make sense at all: r_1 seems to be dangling around somewhat superfluously and only r_2 is relevant for e_2 and also, as expected with a temporal clause, for e_3. e_3 then triggers progression to r_3 which serves as RT for the next sentence. r_1 thus never becomes current RT, and one wonders why it is introduced in the first place. Let's visualize the whole process in a sketch, marking the relevant steps:

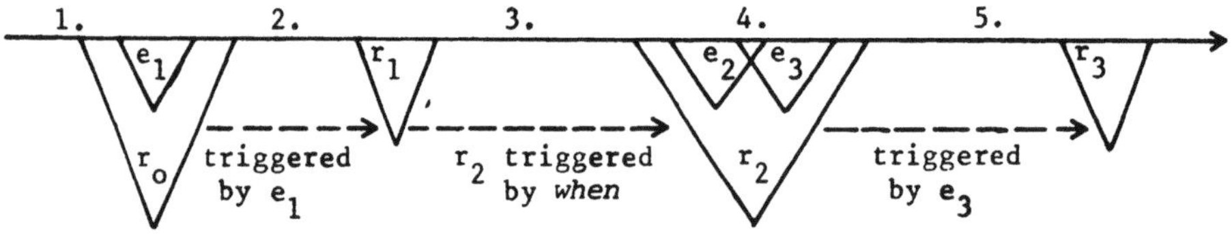

54 Partee (1984:258 f.), example (16) and DRS (16).

With this analysis of (3.22) one might well ask whether updating means the addition of a new RT, while the old one is still there, but forgotten.

Partee then modifies her analysis of *when*-clauses: "... a (preposed) *when*-clause in a linear narrative triggers the introduction of a new reference time located just after the event described in the *when*-clause" (Partee 1984:261)[55]. This implies that in the analysis of (3.22) the steps of e_2- and r_2-introduction and particularly the relations $e_2 \subseteq r_2$ and $r_1 < r_2$ are replaced by (&) $r_1 < e_2 \lesssim r_2$, which is done by Partee in fn. 29, p. 283

Now r_1 has a purpose. It serves to place e_2. But still, the fact remains that e_2 introduces its own RT. There is also the problem that the past tense in the *when*-clause is treated differently than past tense in other clauses: we get $e_2 \lesssim r_2$, when we used to have $e_2 \subseteq r_2$ ($\lesssim$ means "just after").

More disconcerting still are the results we get if we now speculate on the possible function of additional adverbs. Granting that such speculation is always unfair in the absence of some precise definition, I would still like to describe the possibilities so as to show what a precise definition of adverbs should and should not do.

If we assume that adverbs "characterize", i.e. modify, a given RT we still have a choice as to which RT it modifies, and both possibilities create absurdities: Let r_1 be the progression RT passed on by the previous sentence and let r_2 be the one introduced by the *when*-clause, which is, according to (&) to be found just after the *when*-event. If we decide that *at six* modifies r_1 and we follow (&) we have the following situation:

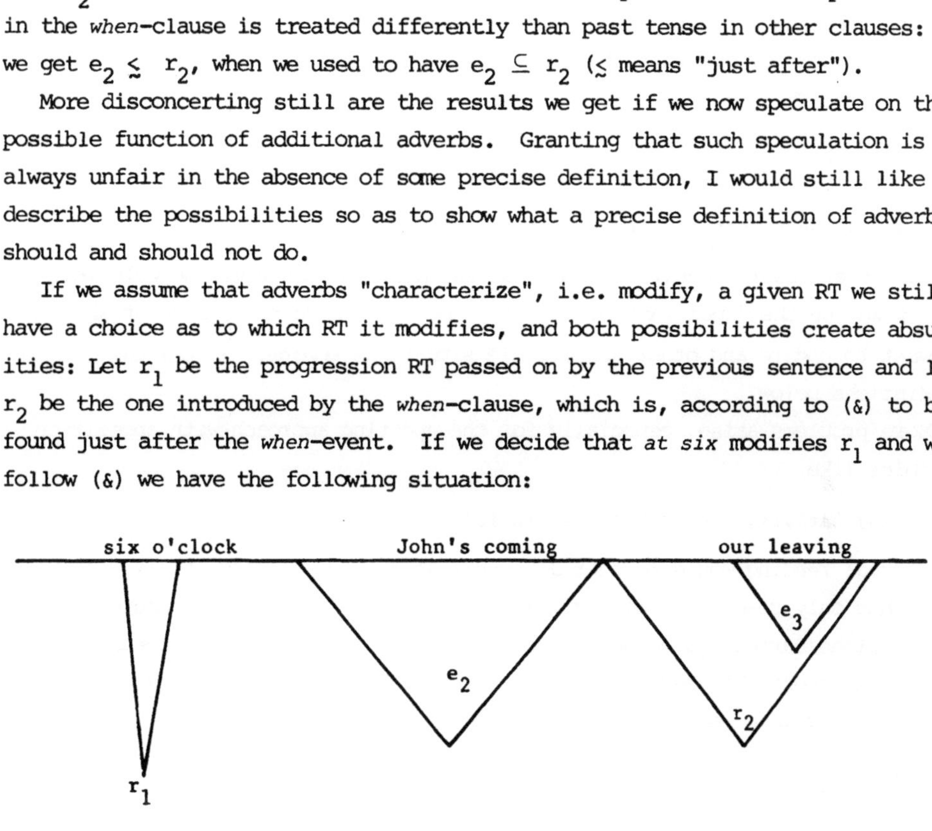

55 Other analyses of *when*-clauses especially simultaneity-*when* are discussed in Hamann: Temporal Clauses, forthcoming.

We therefore leave some time after six, which is a feasible interpretation, but John must also have come after six, which is absurd. We cannot take the progression RT, for this reason and also because of examples like:

(3.24) John came home late. When he awoke in the morning ...

where *in the morning* cannot modify an RT just after John's coming home, cf. (3.21).

If we therefore take r_2 to be the reference time characterized by the adverb, we know that r_2 is just after the *when*-event, i.e. John's coming in (3.23), wherefore John must have come just before six, which is just as absurd as the previous situation.

A possible way out is to have adverbs in temporal clauses modify not RT but event time. Then John came at six, *when* moves RT to just after six and so we left just after six, which is at least feasible. If one takes this way out, however, one gets only one reading for:

(3.25) When John had arrived at six, we left

which is at least undesirable, so that another approach would be welcome.

If we use the "updating" possibility, we again have the choice of which RT we want to update and other undesirable situations emerge, cf. our discussion of Hinrichs below.

More problems arise, especially for the updating approach with apposition examples like

(3.26) At six, when John came, we left.

The normal reading would be that John came at six <u>and</u> we left at six. I suspect that only readings implying that we left after six can be computed with the Hinrichs/Partee system and Partee's analysis of *when*. That something has to be done about this sequence-*when* analysis can also be seen in examples involving *and then*, normally the sign for an already moved RT:

(3.27) ... he smiled at her ... and then, when they were no more than 30 feet up she realizes ... (Higgins)

The problems stem as much from the unusual treatment of *when*-clauses as from the rather vague analysis of adverbs in Partee (1984). The latter can be remedied by checking the formal definitions in Hinrichs.

With regard to reference point, adverbs like *in 1984, on Monday* or *at six* function like this in Hinrichs (1981):

First, there is a scoreboard (cf. Lewis (1979)) for temporal entries, i.e. with space reserved for reference points and for temporal specification like year, month, day etc. As a further restriction, an adverb like *on Monday* may only be introduced into the appropriate space if the "upper" specifications like year, month, week are already filled up. Given such a scoreboard, an adverb introduces two intervals, e_i and e_{i+1}, into the discourse structure, it removes from the scoreboard the given RT and the temporal information of its kind (*on Monday* wipes out the entries in the slot for weekdays and "lower" specifications like time-of-day, not the entries about year or month). e_i, the interval named by the adverb, is written into the appropriate slot, weekday in the above example, and then e_{i+1} is written into the slot for RT with the note in the discourse structure that $e_{i+1} \subseteq e_i$.

This means that adverbs remove any progression RT and the *when*-clause, if taken seriously in its adverbial function, should remove the superfluous r_1 before establishing anything new - which it doesn't, neither in Hinrichs nor in Partee. Adverbs further introduce a frame (e_i) and also the new RT(e_{i+1}) to be placed into this frame. In this sense, the adverb indeed "overrides" the old RT (it wipes it out), it "introduces" the new RT, and it "characterizes" it by specifying the frame. Thus Partee's terminology is clarified though it leaves us in the dark as to what level we are manipulating (scoreboard or discourse representation) when we wipe out and introduce and as to how the scoreboard fits in with the discourse representation. Is it a part of it? Something else becomes clear too: as an adverbial clause, a *when*-clause must introduce its "own" RT.

Hinrichs' analysis now takes care of the step from S_1 to S_2 in (3.21) and (3.24) repeated here for convenience as:

(3.28) John came home late. In the morning he woke up refreshed. (S_1 above "home", S_2 above "he")

and

(3.29) John came home late. When he woke up at six, he felt refreshed.

because wiping out the previous RT avoids any temporal conflict with it.

The nonsense arising from

(3.30) At six, when John came, we left.

is obviously due to the *when*-clause analysis and could be remedied.

The problems of

(3.31) When John came at six, we left

remain, however, though in a different guise. Now, we do not know which RT should be wiped out or, in other words, in what order we should process the sentence. If we wipe out the progression RT, a step which makes sense in view of (3.29), we get nowhere, because the *when*-clause ignores the given RT. To wipe out the *when*-clause RT gives us the paradoxical situation of having left at six (we have replaced the *when*-clause RT by *six o'clock* and this is the RT relevant for the main-clause), without knowing when John came.

Hinrichs expressly abstains from treating double adverb modification, but one might expect that there is some kind obvious treatment that ties in with the system. As it is, the whole system seems to need revision in order to allow for adverbial modification of an already given and probably even specified RT.

Apart from these problems of detail we are forced to return to the old question of whether the adverb or, for that matter, the Simple Past is really the source of the reference point. We are again faced with the problem of explaining where the reference point comes from in cases where the progression RT cannot be used and there is no adverb. Note that the most usual reading of

(3.32) John came home completely exhausted. He woke up refreshed and saw the problem in a new light.

is the one implying a gap of five to eight hours, which cannot be explained by progression to "just after". The gap can be smaller, certainly, John could have had a nap of five minutes or woken up in the middle of the night, but the point is that there is - apart from context - nothing to support the one or the other reading.

Therefore we probably should welcome the relative vagueness of a pragmatic principle like Nerbonne's. The reference point for waking up is assumed to be definite (which is correct in this case) but it is vague in the sense that we only know it is somewhere after John's coming home, ready for specification by an adverb or further context.

Another possible model comes to mind when we think of Hinrichs' scoreboard reserving space for entries concerning the reference time. Ritchie (1979) following Isard (1974) and using computer terminology speaks of a register or slot reserved for the temporary storage of items, here reference time. Therefore we could have three different kinds of activities we perform when using temporal expressions, all concerned with reference time:

1. We could have commands establishing a slot (not its contents) which might be what the progression is about, it establishes a new slot,
2. we could have commands filling the slot, specifying the content more or less precisely, and
3. we could have commands finding the slot, reading its contents and probably transferring them to other slots.

If the past tense turns out to be a search instruction for RT (a rule of type 3), the adverb only specifies the contents of the RT-slot (rule of type 2), and progression turns out to be too special to set up the slot in all discourse situations, we will have to look elsewhere for the origin of RT or its slot.

The investigation of the presupposition of existence of individuals in definite descriptions might show parallels to the use of temporal expressions. It might turn out that we not only presuppose the occurrence of events but also their reference times. In this case the reference time slot would be set up automatically as soon as a sentence is tensed or contains other types of temporal expressions. Progression in narration would then be one type of this automatic procedure and the descriptive mode would obviously require another.

Clearly, much remains to be done on the concept of reference time.

TIME ADVERBIALS IN ENGLISH AND REFERENCE TIME

Janet Harkness

1. INTRODUCTION

In this article we set out to consider the role time adverbials (TA) in English play in temporal reference. After indicating the criteria by which we assign expressions to the category of time adverbial or not, we distinguish between the times contained in TA and the times named by TA, which latter we call TA time. We then discuss TA in relation to reference time (RT); we distinguish between RT and the time TA refer to and demonstrate that where TA are not required to characterize RT, they often characterize ET or some other time/event. Finally we examine TA in combination with present perfect and pluperfect forms and consider the relevance of TA for deciding whether temporal reference is definite or indefinite.

2. WHAT ARE TA?

TA, essentially, are adverbs and adverbials which refer to temporal locations or, as it is often put, to times on the time line. Thus if we conceive of time as a line running from left to right, and mark the moment of speech as ST (speech time) on this line:

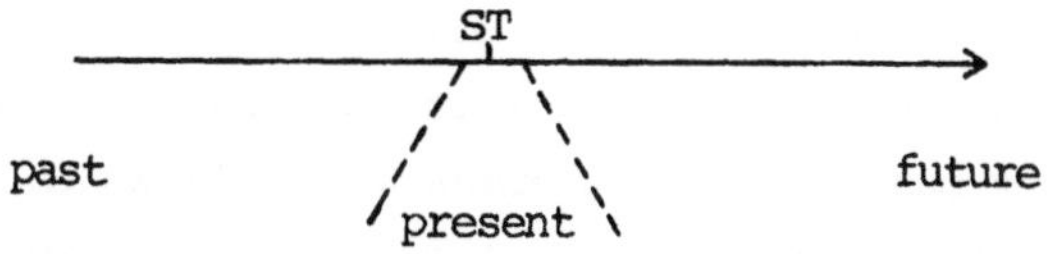

any times left of ST are to be thought of as times anterior to ST (past), any times to the right of ST as posterior to ST (future) and any times coinciding with ST as simultaneous (present time). The precise manner in which TA refer to times and the character of the times referred to can vary considerably, as the following examples illustrate. TA such as *on the 8th of May 1919* name times of fixed identity; TA such as *tomorrow*, on the other hand, name times whose identity is established on the basis of whatever time constitutes 'now'. TA such as *at 5 o'clock* and *at Christmas* are essentially incomplete, since they refer to a time only by virtue of it being located via a set of cyclically returning entities (such as festivities or clock times). The particular clock time or festivity referred to is determined by the larger context.[1] TA such as *earlier* and *later*, on the other hand, refer to times by indicating the temporal relation involved between the time referred to and a contextually given time.

The temporal profile of events

TA inform us about temporal locations. Our interest in temporal locations, at least in language, is closely connected with our interest in communicating information about events. The temporal location of an event, however, is only part of the information relevant to what we will call the temporal profile of an event. Knowing, for example, that a certain event took place on a certain day does not inform us about the duration of the event, the nature of its temporal extension (whether continuous, spasmodic, etc.), whether this occurrence of the event was one of regular occurrences, occurring frequently or infrequently as the case may be, nor whether the occurrence was long-awaited or happened unexpectedly.

TA cannot provide us with such information. There are, however, a number of expressions in English which do provide such information and, in combination with the tense chosen, aspect used and event notion involved, such expressions characterize the temporal profile of the event.

The following examples help illustrate these points; the expressions *after all*, *long-awaited*, *suddenly* and *again* all contribute to our knowledge of the temporal profile of any event they are used to describe. We learn, for example, that previous to enactment there was doubt as to whether the event would take place at all (*after all*), that there was a gap considered as lengthy between the announcement of future enactment and actual enactment (*long-awaited*), or that the occurrence referred to is not the first occurrence (*again*). Although clearly distinct from information about the temporal location of an event, such information is nevertheless related to the conditions, nature and timing of its occurrence.

---

1 Bull (1960:10) notes of calendar references (applicable to clock time references) that they are: "utterly useless for any practical purpose unless all their time intervals and their finite interval series are axis bound in terms of a second axis of orientation, that is, an event inside the person who wants to use the calendar." Like Lyons (1977:684) Bull clearly views such expressions as ultimately connected with deixis.

Duration and frequency

Two categories of expression whose primary role it is to provide information about the temporal profile of events are durative expressions and frequency expressions. Durative expressions indicate the duration of events in various ways, frequency expressions provide us with information about the number of occurrences of an event or the relative frequency of occurrence. Although information of this sort cannot itself allow us to identify the temporal location of events, under certain conditions durative expressions, and in some cases frequency expressions, can help locate an event on the time line (cf. Bennett/Partee (1978:22f.), Nerbonne (1983:86f.)). A brief consideration of each type of expression will indicate how this is possible.

Frequency expressions

The primary information carried by frequency expressions is information about the repetition of occurrence of an event or events or the absence of repetition. There are various ways in which repetition can be indicated. First, however, a distinction must be drawn between enactment and frequency. Thus *never* cannot qualify as a frequency score, since it indicates zero enactment.[2] *Never again*, in contrast, indicates the absence of repetition beyond some contextually indicated occurrence(s). *Once only* indicates zero repetition, expressions such as *twice* indicate the concrete number of occurrences involved, and expressions like *often* entail repetition and convey information about the (relative) degree of frequency. *Often* is context-dependent and as such is only fully interpretable in a definite (tensed) situation. Thus we know approximately what *often* conveys but require it to be used in context to fully appreciate its meaning. Without a context there is no real way of recognizing quite what *often* involves.

Frequency expressions such as *regularly*, *occasionally*, and *sporadically* convey information about the distribution of occurrences, uneven or regular scatter; expressions such as *constantly*, if combined with repeatable events, signal repetition. With states and processes, provided these can be set limits, *constantly* can indicate uninterrupted continuance or repeated instances of the state of process.

2 This is not to deny that *never* is a universal quantifier expressing a negation of *always* and *sometimes*.

Not all events can be assigned frequency in a meaningful sense. In some cases the limits on repetition are of more pragmatic nature (< Lily die >), in other cases the event notion itself excludes repetition (< Lily find out (about the accident)>).

Since all events which have occurred have occurred at least once, as said, the fact of enactment (which entails an occurrence) does not automatically permit the assignment of a frequency score of 'once'. In essence,a frequency score can only be assigned to an event which potentially can be repeated; such events require that we be able to envisage further occurrences of the same event.

Frequency expressions and temporal location

If occurrences of an event are known to be located within a given span of time, certain frequency expressions can provide information about the temporal location of individual occurrences. Thus in:

Last week I saw John every day

every day in combination with the TA *last week* enables us to locate at least one occurrence of <see John> within each day of the week indicated by the TA. In a similar fashion, provided we know which span of time is being referred to, the sentence:

The baby was fed at hourly intervals

allows us to locate at least one feed within each hour of this time. The instruction on a baby food packet "Feed at hourly intervals", in contrast, cannot permit location on the time line since reference is not made to a tensed situation.

Durative expressions

Durative expressions modify events and states by providing information about their duration; they are not time-orienting in function. Thus although the duration of an event can be thought of as its extension in time, in order to be able to locate the event (in its extension) in time, we require to know times as well as extensions.

Certain expressions indicate extension by reference to points on some time-measuring scale. Expressions such as *from 1 to 6 o'clock, from the beginning of May to the end of June, from Christmas Day to New Year's Eve* can indicate extension (duration) by indicating that it is equal to the span between the two points/times mentioned, that is, five hours or two full months, etc. Other durative expressions state concretely the duration involved: *for three hours, within 5 minutes. Within*-expressions state the maximum possible or estimated duration.[3]

Durative expressions with times as anchors

Using the example of a note pinned to a door saying "Back in two hours", Fillmore (1975:38) points out that certain expressions, such as *(with)in an hour*, cannot allow us to identify times without there being a time from which to count forward, in this case, to count forward an hour. Similarly, a note saying "Here between 5 and 6 " cannot be fully interpreted (apart from habitual readings) unless we know which day is referred to (and perhaps whether the speaker is using a 24-hour clock or a.m. and p.m.). Given a temporal anchor, provided either concretely or contextually, the durative expressions in:

I'll leave in three hours (from now)
I'll be back in an hour (from now)
I have to stay here for another three hours

---

3 Nerbonne (1983:93f.) calls these Frist adverbials, using a German word which he paraphrases here as "the term within which something happens". Nerbonne's formal analysis of expressions such as *in einer Stunde* ('in an hour'), following Dowty (1979:333), requires that propositions be true of the interval referred to but not true at any sub-interval within it. Note that it is possible to distinguish between *in an hour* and *within an hour* whether anchored to a time or not). *Within*-expressions indicate the maximum interval: Thus:

He wrote a poem in an hour

suggests he took an hour to complete writing the poem.

He wrote a poem within an hour

can also be taken as implicating he took less than an hour (cf. Harkness (1985:370f.)).
In either case, the durative expression can only help indicate a temporal location if it itself can be related to a time on the time line. Where this is the case, we can speak of the durative expression having a time line 'anchor', by which we mean it is related or tied to the time line by means of a time which functions as its anchor. (The term is taken from Smith (1978:47ff.) who uses it in connexion with TA.)

as well as the written message:

11.15: Tuesday 3/12/86: Back in an hour

allow us to compute the temporal location of the event referred to.

In a later section we consider TA which require to be anchored to a time in order to be interpreted fully. We distinguish between such anchors and count times, the latter being simply times from which we count. These may but need not be anchors. The times required to interpret durative expressions beyond their durative content are first and foremost time anchors. In contrast, "closing time" in:

Back within an hour of closing time

and in the TA in:

Back shortly after closing time

is a count time. Without an anchor (a time location) for the whole expression, we are unable to identify the time intended for return and know only that it is within a short span of time after some unlocated closing time.

For-expressions

One of the most frequent types of durative expression combines *for*, indicating duration, with another element indicating extension (*3 hours*, etc.). Instead of a time quantity, some *for*-expressions involve events which are tied to a particular time of year and which in part at least are characterized by being tied to this time of year; examples are *for Easter*, *for the summer holidays*, etc. Provided with an anchor, such expressions are commonly used to indicate times rather than simply durations. Thus:

Lily is coming for Easter

is as likely to be used as an indication of when Lily will come to visit as it is to indicate merely the duration of her stay. In a tensed situation in fact, the durative, in contrast to *at Easter*, automatically provides both pieces of information.

Distinct from these are a number of TA which combine *for* indicating duration with an expression which refers to a time on the time line (rather than a calendar reference). The majority of these TA are anchored to ST, examples being *for the present, for now, for the time being.* In the case of *for the past six months*, the TA time begins six months anterior to ST and extends up to ST. It is worth noting that the presence of "past" here automatically identifies the expression as a TA, which refers to time beginning anterior to ST. *For the last six months*, in contrast, can be used to refer to times if anchored, but can also simply indicate duration, as in:

The course usually spends three months on theory and for the last six months concentrates on application.

For-X durative expressions and *since*-X TA

A number of parallels can be noted between *since*-X TA (*since Christmas*) and *for*-X expressions. The durative expression in the following sentence:

I have lived in Britain for 6 months

if understood as anchored to ST (a salient although not required reading in combination with the present perfect), allows us to compute a t_i (initiation time) of the state <live in Britain>.[4]

In the reverse fashion:

I have lived in Britain since last Christmas

allows us to compute the duration of <live in Britain> as that period of time between the Christmas before ST and ST (cf. Schopf (1984:346), Harkness (1985: 301ff.)).

Jessen (1974:555) notes that both *since*-expressions and *for*-expressions can have a "universal" and an "existential" reading. The existential reading for:

Fred has lived in Paris since 1950

allows that Fred has lived in Paris several times within the span of time indicated between 1950 and ST. The existential reading for:

---

4 We distinguish between 'events' on the one hand (achievements, accomplishments, punctual changes of various kinds; cf. Vendler (1957) and Schopf (1984:Chapter 2)) and 'states' and 'processes' on the other. We use 'event notion' for all three when we wish to refer to the temporal structure involved in a given event, state or process.

Fred has lived in Paris for 10 years

is that in which *for 10 years* indicates only the duration of <Fred live in Paris>.

Already, still, yet

The adverbs *already, still* and *yet* and their non-positive correlates *not yet, no longer* (and *not anymore*, etc.) are sometimes treated as TA (Schopf (1984: 142-159), Harkness (1985:206-272) and, in essence, Jessen (1974:570ff.)), and sometimes as markers of aspect (Traugott and Waterhouse (1969)).

Since, for example, in both:

Lily has gone

and:

Lily has already gone

the event <Lily go> is anterior to ST, it has been suggested that these adverbs are redundant as far as any time-orienting function is concerned; in both cases the present perfect tense indicates that ET is anterior to ST and, following Reichenbach (1947:297), RT coincides with ST. The sentence with *already* does not provide any more information about the location of ET. In a comparable situation with the simple past tense, on the other hand:

Lily left/went last Tuesday

the TA time provides temporal coordinates for the RT, which includes the ET, already known to be anterior to ST by virtue of the tense used.

Already, still, yet,etc., do not establish TA times within which ET is to be located. Instead they indicate that the given arrangement of ET, RT, ST stands in contrast to a different arrangement *(already, (not) yet)* or they relate the present arrangement to an earlier arrangement which it either matches (*still*) or to which it stands in contrast *(no longer)*. Jessen (1974: 610ff.) notes differences in the event notions which collocate with each; *already* and *(not) yet* require events which can 'begin', that is, in Jessen's terms have an "inceptive" core, *still* and *no longer* require events which can 'terminate', that is, have a "cessative" core.

These adverbials thus constitute a special case among TA; rather than providing co-ordinates for RT/ET or ET, they provide additional information in the form of entailments or implicatures about other potential past or future arrangements of ET, RT, and ST. Thus the sentences in which they occur refer both to the given configuration of ST, RT and ET and to a further configuration of these (cf. Jessen (1974:618f.) and Schopf (1984:145ff.)).

3. A CLOSER LOOK AT TA

The first step in our consideration of the temporal reference of TA is to distinguish between the times TA often contain and TA times, that is, between the times named in TA and the times named by TA. When we speak of times in the following, we intend these to be understood as intervals of various kinds. Moments of time should thus be thought of as minimal intervals, stretches of time as non-minimal intervals (cf. Bennett and Partee (1978:11f.), Kamp (1979)). A number of TA require that we distinguish further between intervals which involve an indivisible span of time, which we will call filled intervals, and and those which do not. All minimal intervals are to be understood as indivisible, as are the non-minimal intervals involved in TA such as *nowadays, these days, throughout 1919,* and *all through that night.*

The extension of the times involved in TA

A glance at only a very few TA reveals that the extension of times involved in TA and the way these are referred to can vary greatly, since clock divisions, calendar and other publically or privately recognized temporal divisions, subjectively or objectively assessed extension of the temporal relation involved (*ages ago, three weeks ago*), and the so-called deictic expressions (*today, tomorrow*) must all be considered. Many of these, as we shall see, are essentially incomplete. We have already mentioned, for example, that references to points in cyclic systems of dividing time always require to be tied to a time location in order to be interpreted properly.

The clock time TA *at five o'clock* contains what is understood as a minimal interval, identified as the clock time named. The year date TA *in 1919* contains a stretch of 12 months. *Today* (single day) contains a stretch of

24 hours, *so far* involves a span of time stretching from one (contextually or otherwise) established time to ST, or RT , if RT does not include ST. *Nowadays* consists of a filled interval which includes ST and a considerable 'extended present'.[5] We consider some of these TA in greater detail later.

Some TA do not contain times, although they do have a TA time, that is, refer to times. The TA *later* and *earlier*, for example, contain only the temporal relationships of posteriority and anteriority, respectively. In contrast to these, *soon* also quantifies the degree of posteriority involved (cf. Schopf (1984:131-133), Åqvist et al (1977)).

The times in TA and TA times

For TA which contain a time and involve the relationship of simultaneity, the time contained in the TA is also the TA time. Examples of various kinds of TA in which this is the case are *in 1919, at five o'clock , at Christmas, over the weekend, nowadays* and *these days*.

With TA which involve a time and a temporal relationship other than simultaneity, the TA time is necessarily different from the time contained in the TA; examples are *after 5 o'clock, before 1919*.

In the case of *until 1919*, the year 1919 alone cannot constitute the TA time. The TA time is a filled interval beginning anterior to 1919 and finishing at 1919. *By 1919* is often taken to imply reference to time anterior to the year named as well as the time named (cf. Harkness 1985: 293-295). Thus in:

By 1919 they had found a solution

in contrast to:

In 1919 they had found a solution

<they find a solution> is located within a period of time which in its last phase includes (some or all of) 1919. Hence the ET could also be anterior

5 For our present purposes we can ignore the philosophical and physical problems in defining present time and content ourselves with Jespersen's comments: "Theoretically it is a point, which has no duration, anymore than a point in theoretic geometry has dimension. The present moment, "now", is nothing but the ever-fleeting boundary between the past and the future... But in practice "now" means a time with an appreciable duration, the length of which varies greatly according to the circumstances..." (Jespersen (1951:258)).

to the year named. With *in 1919*, the TA time does not extend beyond the year named. We see later that the pluperfect here means that TA time does not necessarily provide a setting for ET.

Anchor-dependent and independent TA

A distinction can be drawn between TA which refer to times of unchanging identity and those which refer to times whose identity is determined by the axis of orientation, that is, first and foremost , the hic et nunc of the speaker. Independent or absolute TA, as they are sometimes called (cf. Fabricius-Hansen (1984:207), "absolute rahmenbildende Betrachtzeitadverbiale"; Bennett/Partee (1978:22), "non-indexicals"), have TA times of fixed identity on the time line; they consist of dates, historical events, the names of eras, epochs, etc. Thus the year named by a TA such as *in 1919* is always the same year. The year named by the dependent TA *last year*, in contrast, changes in accordance with the context of utterance. The relation between times of fixed identity and ST changes constantly, the relation between times whose identity is defined by the position of ST and ST remains constant.

The great majority of TA are dependent, that is, they require to be fixed or anchored to a time established on the time line (or able to be established on the time line) before their TA time can be fully interpreted.

Types of dependent TA

Any TA which do not involve times of fixed identity are dependent TA; within this category a number of distinctions can be made. Some dependent TA consist solely of a relation (*afterwards*), others qualify this relation (*soon*), others involve special names for times, such as *today, yesterday, this week*. (Most of these latter have exact non ST-anchored equivalents: *that day, the day before, that week*.) Finally, there are expressions involving reference to cyclically organized events of the year or clock *(at Christmas, after Easter, at 5 o'clock)*.

Expressions such as *during the concert* can be located on the time line via the unique event referred to here with "concert" and are thus not anchored in the sense just outlined. They can, we suggest, be more readily compared to *during 1919* (but see Fabricius-Hansen (1984:20)).

Different anchors

Schopf (1984:123f.) distinguishes between dependent TA anchored to ST ("sprechzeitgebundene"), *at present*, those anchored to an RT ("referenzzeitgebundene"), *the day before*, and those which can anchor to any time, whether ST or not ("ungebundene"), *later*. The intention was, we presume, to distinguish between dependent TA which can anchor to ST, those which must anchor to ST and those which cannot anchor to ST. However, calling TA which cannot anchor to ST "RT-anchored" can be misleading, since for some TA at least the RT will include ST (*nowadays*). When necessary, therefore, we propose to distinguish between TA which anchor to ST as "ST-anchored", those which cannot anchor to ST but do require an anchor as "non-ST-anchored" and those which simply require an anchor whether this be ST or another time as "T-anchored" (time-anchored).[6] *Some time ago* is then ST-anchored, *some time before*, non-ST-anchored, *soon*, T-anchored (but see Levinson (1983:74)), and *three weeks ago yesterday*, a ST-anchored TA in which 'yesterday' is used as a count time from which to count off three days. The whole expression is anchored, as said, to ST.[7]

The temporal relations possible in TA

Three temporal relations are possible between entities and times or times and times; posteriority, anteriority and simultaneity. In that one event can overlap with another, one time include or partly include another, these relationships are not mutually exclusive. Dependent and independent

---

6 For the purposes of this article we have left aside the question of whether ST-anchored TA are to be considered as merely this (and what then can constitute an ST) or as truly deictic. Fabricius-Hansen (1984:216) takes the former view, Lyons (1977:678f.), the latter. Many of the distinctions relevant to temporal deixis are discussed in Rauh (1984a and 1984b).

7 Fabricius-Hansen (1984:40f.), following Bäuerle (1979), uses the term "evaluation time" (Evaluationszeit) for anchor times. Unfortunately she and Bäuerle also use it for count times, that is, for times from which we count off quantities of time, etc. (cf. Fabricius-Hansen (1984:208f.)). As mentioned in connexion with durative expressions, we consider it important to distinguish between the two: an anchor can always function as a count time *(three days ago)* but the count time does not always function as an anchor, viz. 'yesterday' in *three days ago yesterday*.

TA alike involve these three relationships. For independent TA the relationship holds first between the time of fixed identity and the TA time, second, as determined by ST, between TA time and ST. For dependent TA it holds first between the TA time and the anchor time, second between times contained in the TA and the relationships expressed (*at Christmas, after Easter,* etc.). Thus *after 1919, soon* and *tomorrow* involve posteriority, *after 1919, earlier* and *yesterday* involve anteriority and *in 1919, at present* and *today* involve simultaneity.

TA times, temporal relationships and times referred to

It will be evident that only in the case of TA consisting of a time and the relationship of simultaneity will the TA time be the same as the time contained in the TA (*in 1919, nowadays*). Where a time and the relationship of anteriority or posteriority are involved, the TA time will necessarily differ from the time contained in the TA (*after Christmas*). For these TA, the time in the TA has a similar function to the count time 'yesterday' in *three weeks ago yesterday*, in that it serves as a time from which to apply the relationship. In the case of *tomorrow* and *yesterday*, the relationship of posteriority and anteriority respectively holds between the anchor (ST) and the day named by the TA. In a TA such as *after tomorrow* 'tomorrow' is anchored to ST and the TA time is in turn posterior to this time (but see Bäuerle (1979:122f.)).

The times to which TA can refer seem therefore fairly straightforward and we might want to propose the following rules: if the relationship involved is not simultaneity, take the time named, apply the relationship given and thus posit the TA time. If the relationship involved is simultaneity, the TA time will be the time named. However, there are situations in which at first glance these rules do not appear to apply and we consider such contexts shortly.

4. THE ROLE OF TA IN TEMPORAL REFERENCE

In the sense that naming a time can be understood as referring to this time, TA can be said to refer to their TA time. The time referred to in a given utterance or piece of discourse, on the other hand, is RT. Reichenbach

(1947) who uses S, R and E rather than ST, RT and ET, does not distinguish between TA time and RT. His examples, indeed, suggest that TA identify RT.

In recent years, work in the field of temporal reference (Bennett/Partee (1978), Smith (1978, 1980, 1981), Bäuerle (1979), Nerbonne (1983) and Fabricius-Hansen (1984)) has raised questions which make the need to distinguish between the two clear. At the same time, the role played by TA in temporal reference is closely connected with RT and it is to the connexions between the two that we now turn.

TA and RT

Partee (1973) and Nerbonne (1983) have pointed out that simple past tense sentences are likely to be read as referring to definite times. To take an example from Nerbonne, the time referred to in:

Cal forgot to turn out the stove

is the occasion on which, it is asserted, Cal forgot to turn off the stove. Thus we do not need TA in order to refer to definite times. As it stands, however, the sentence gives no indication of when the definite time referred to is located on the time line other than that it is anterior to ST. Thus the TA in:

Cal forgot to turn out the stove last Tuesday

does not indicate that we have a definite RT but rather provides information about this RT anterior to ST.

Reichenbach (1947:288) speaks of R as <u>the point of reference</u>. The reference point can coincide with S, E, or both, or be distinct from both.[8] Reichenbach's discussion of R indicates that, at least in part, it is to be understood as a conceptual reality rather than a time located on the time line in the sense that ST and ET are times on the time line. R is thus on the one hand the point from which "events are seen" (1947:288), a vantage point which provides the key to Reichenbach's neat distinction between the simple past and the present perfect ((1947:289 and 297), see, too, Hamann (this volume)).

---

8 When referring here to Reichenbach , we use R, S, E; when referring to what we consider is both what Reichenbach intended and what we intend with RT, ST, ET, we give both (R/RT), later returning only to RT, ST, ET.

On the other hand, particularly when TA are involved, we frequently find that this conceptual vantage point takes on rather solid form, since it is characterized as located over or within a TA time.

Were this only the case when R/RT coincides with E/ET or S/ST, one might argue that it is the latter which are characterized. This in itself would be problematic since *nowadays*, for example, hardly only refers to ST even if we understand ST loosely (see, too, Hamann (this volume)).

However, as we saw in one reading of:

Lily had spoken to Thomas at 5 o'clock

the TA can characterize the RT independent of ET and ST. And, as Hamann (this volume) points out, RT can be relevant for the truth conditions of a sentence. We suggest therefore that RT should be seen, as appropriate, both as a vantage point and as a time.

Keeping this modified notion of RT in mind, with a Reichenbachian analysis of simple past tense sentences, ET coincides with or falls within the time of RT.

Although, as we shall see, this does not hold for all simple past tense sentences, in the following:

Cal forgot to turn out the stove last Tuesday

the ET is either equal to RT or lies within RT and the RT lies within the TA time. If, for example, both speakers know that they are referring to the evening of the past Tuesday most proximal to ST, when Cal cooked dinner, RT is this occasion on which Cal forgot to turn off the stove. The TA time provides a temporal setting within which RT lies.

When a TA is present in a simple past tense sentence and is used to characterize RT, with a few exceptions the RT will be a definite time equal to or lying within the TA time. This does not apply, however, if the TA characterizes a time other than RT (in which case RT would be definite but outwith TA time) or if the TA is one of the few which signal indefinite reference, such as *sometime* or, for future events , *one day*.

As Nerbonne (1983:25-27) , for example, notes, the fact that a time is definite is independent of whether information about its location on the time line is available or not. And, although a TA may only provide an approximate setting and the details of RT and ET are left open, it would be wrong to think of these times as indefinite, they are simply unspecified (but see Bäuerle (1979:45) on event time). In contrast to the above, the temporal reference in the sentence:

Cal has forgotten to turn off the stove upon occasion

is indefinite. This is directly attributable to *upon occasion*, since:

Cal has forgotten to turn off the stove

can refer to a definite time (see discussion of the present perfect later and Matthews (this volume)).

Although there are often important differences in the approach taken and in the underlying conceptions of RT and TA, a number of authors in recent years have suggested that TA in some fashion provide temporal settings; Bäuerle (1979) and Fabricius-Hansen (1984) have them specify exactly the time "looked at" in a sentence, Bennett/Partee (1978) and Nerbonne (1983) speak of certain TA as "frames", Schopf (1984), as times within which or over which RT is located.

We consider it essential to distinguish between the temporal setting TA can provide and RT, ET and ST. TA refer to TA time and TA time may characterize RT, ST, ET, or a combination of these. RT can be equal to TA time but is often not. The information the TA provide can be extremely concrete (time, day, month and year named) but it can also be as subjective and context-dependent as *ages ago*.

TA and "Betrachtzeit"

The different approach taken by Bäuerle (1979) and Fabricius-Hansen (1984) is a direct consequence of the fact that they dispense with RT and are thus left with ST, ET, and the time given by TA. These, they suggest, can specify the time referred to or "looked at". Bäuerle distinguishes three parameters (1979:47, my translation):

Sprechzeit/speech time	:	time of utterance
Betrachtzeit/time looked at	:	time referred to, looked at, often specified by temporal adverb e.g. the day before today ('yesterday') or the 7.1.49 ('on the 7.1.49').
Aktzeit/event time	:	time at which the described event(s) take(s) place, a sub-interval of the Betrachtzeit.

Bäuerle points out that his Betrachtzeit/time looked at (hereafter BZ) does not have the standpoint function which he sees as the defining feature of Reichenbach's R (1979:48). Bäuerle discusses R as a standpoint from which events are described despite the fact that the speaker is actually at ST and not at the time used as a standpoint. This clearly differs from our understanding of R (cf. Hamann and Schopf (both this volume)).

In addition to BZ, Bäuerle introduces an evaluation time (Evaluationszeit), which is to be understood as "the time counted from". Evaluation time (EVT) can be ST (as it would be in *a week ago*) but can also be another time. Bäuerle (1979:50) demonstrates this with the example *around noon a week ago yesterday*, in which he suggests ST is only the EVT for the first step, 'yesterday' is the EVT for the next step and a day 7 days before 'yesterday' is the EVT for the last step from which 'around noon' is counted off. Thus in the first step Bäuerle's EVT is both an anchor which ties the dependent TA to the time line and a count time from which times are counted off. The other EVT are not anchors, and in any case Bäuerle does not distinguish between these two. Bäuerle then concludes that Reichenbach's R is his EVT. Quite apart from the problems this raises for the analysis of various tenses (cf. Hamann and Schopf (both this volume)), it leaves us without an anchor for dependent TA. To regulate the appointment of TA such as *a week ago* in:

We went to the cinema a week ago

to ST, and in:

We went to the cinema a week ago yesterday

to the time of 'yesterday', Bäuerle also introduces the notion of ST-relative (sprechzeitrelative) "shifters". He explains their function as follows (t_o = ST, the formulae are replaced by their TA equivalents, Adv_{SB} is used for an adverb which, depending on context, can change its EVT to either ST or BZ):

> The interpretation of a BZ adverb only produces the right results if the Adv_{SB} stands within the scope of an ST-relative "shifter" such as 'yesterday'.... 'Yesterday' pushes the time from which ... 'a week ago' is to be counted onto the day before t_o (in *a week ago yesterday*) and in *a week ago* it is pushed onto t_o (1979:123, my translation).

This seems a rather cumbersome way of reflecting that *yesterday* is anchored to ST and there is no discussion of what determines when or why shifting operations occur.

Of prime importance here is that we note that the TA give the dimensions of BZ, they are not a setting for various times. Hamann (this volume) discusses the problems which result from dispensing with RT and only having a BZ. It is really within the context of preferring BZ to RT that a discussion of the functions Bäuerle and Fabricius-Hansen give TA should lie.

We note, however, that since TA specify BZ, if the TA time is greater than the ET, there is no characterization of ET. Bäuerle makes it a sub-interval of BZ, which means that ET will never lie outwith TA time, clearly contrary to the facts we present (see, too, Hamann (this volume)). He gets at the ET within BZ by assigning a frequency score of 'at least once' to the event (1979: 47). Thus discussing a sentence of Cresswell's (1973): ((John, sleep), yesterday), he says: "the outcome of the analysis must be that at some time or another during yesterday John slept - at precisely the <u>event time</u>: therefore speech time and Betrachtzeit are "definite", the event time remains indefinite within the frame set by Betrachtzeit" (1979:45, my translation).

In the final section where we discuss definite and indefinite reference, we take the view that ET within definite TA times are definite times which are not further characterized. This does not mean that they are particular nor that the speaker knows their location.

Fabricius-Hansen follows the general framework suggested by Bäuerle. She explores the temporal strucutre of TA in greater detail than Bäuerle does and her analyses of individual items, even within the BZ framework contrary to ideas presented here, are stimulating. In paying more attention to the temporal structure of TA, event notions and narrative modes, Fabricius-Hansen also runs into a number of problems Bäuerle does not.

Restricting ourselves to TA, she does not assign a frequency of 'once' to events and to get at ET has to elaborate on the notion of BZ. On the one hand TA time give BZ; thus *nowadays* provides an indivisible BZ. Here, however, she notes that the BZ interval is "vague" but its core "definite" (1984:40). A TA such as *today*, on the other hand, may be used to refer to a BZ which does not have the extension of TA time. Thus in:

Arnim cut his finger today

in which Bäuerle takes the time of 'today' as BZ and has the past tense operator single out that interval anterior to ST (1979:53), Fabricius-Hansen suggests that in such contexts only that part of 'today' anterior to ST is referred to (cf. Fabricius-Hansen (1984:63)). This is tantamount to suggesting that the referring expression refers to less than what it refers to.

One can see why Fabricius-Hansen makes this proposal: since the TA gives BZ and not a setting for BZ, it is impossible for BZ to be equal to TA time since TA time includes time posterior to ST clearly not relevant for <Arnim cut his finger>.

Since we only require TA times to be settings, tense operations, aspectual considerations, the dictates of the event notion and pragmatic consideration can interact to determine whether the time characterized is equal to or lies within TA time. Since ET is not necessarily a sub-interval of the time given by the TA, we can characterize ET without RT and RT without ET.

Why do TA characterize RT?

The foremost role TA have in temporal reference is to characterize RT. As soon as we talk about events that have happened or we envisage as going to happen or which we set up as "real" events (in a narrative, for example), we automatically require times. Times have extension without events, events can only take on form in time. Hypothetical events also require times within the framework of a world in which they have existence. The need for times is reflected in the fact that in beginning a story which opens without any indication of temporal setting, we seem simply to create one in our heads and from here on begin to organize the events and times we are confronted with in the narrative (cf. Couper-Kuhlen (this volume)). The factors which determine whether it is appropriate in a given context to characterize RT are manifold. They depend largely on the type of oral or written text involved and on the details of the particular narrative as well as on the concerns of the teller (cf. Couper-Kuhlen (this volume)).

The tense and aspect chosen in a sentence indicate in most cases the relation of RT to ST, that is, whether the time referred to is anterior or posterior to or overlaps with ST. Information beyond this can be provided by TA. The key question, however, is what determines whether further information about RT (or in some cases ET) is required. It is clearly not determined by the absence of a concrete location alone: in one context we may already know where RT is and nevertheless be given a further characterization. In another context, we may simply have posited RT and have no real information about it and never receive any either (cf. Couper-Kuhlen (this volume), Schopf (1984:377ff.) and Partee (1984)). This is not, however, a question which can be answered here.

When TA characterize ET and not RT

In what might be seen as the negative correlate to our proposal that TA characterize ET if not required to characterize RT, Smith (1981:222) suggests that: "it is a characteristic of English that an ET other than RT is specified only by sentences that have an RT specified otherwise". Smith (1978, 1980, 1981) considers examples in which TA time does not characterize RT, or rather what Smith considers RT. She concentrates on complex sentences and on temporal reference across sentences. Many of her proposals, therefore, lie beyond the scope of the present discussion in which we attempt to establish the fundamental potential of TA and thus restrict ourselves to simple sentences. A further difficulty is that although Smith states she takes Reichenbach as her starting point (1978:43f.), her conception of what RT is in a given sentence does not always tally with Reichenbach's. Sometimes she uses R/RT and sometimes she says a count time (which may or may not be an anchor) is RT. Her interpretation of TA differs accordingly, despite the many interesting observations she makes. Thus whereas we agree with Smith that sentences have one TA, however complex (1981:215), agree that the anchoring potential of TA is important (1981:213f.), can appreciate why she feels sentences without a TA are "incomplete" (1981:217), we do not agree with her analysis of certain structures in a given context. Thus in:

Roger called before noon

Smith (1981:216) says that the tense and 'noon' "specify" a past RT and that 'before' indicates ET precedes RT. In other words, Smith takes RT as specified by the clock time. Furthermore, she frequently speaks of RT being "specified" where RT (or whatever) is only given a temporal setting. And in:

The boys had already eaten dinner (1978:45)

one of a number of examples in which for Smith tense and TA combine to "specify" RT, we would suggest that there is not even a setting given, beyond the information provided by tense. *Already*, as indicated earlier, tells us nothing about this RT, it contrasts it with another potential RT later.

Smith (1981) discusses re-anchoring of ST-anchored TA in complex sentences. Since this brings important characteristics of these TA to light, we outline briefly one or two of her examples. Smith in fact discusses two main kinds of re-anchoring situation, one in which a "sharing" relation obtains, the other in which an "orientation" relation obtains. It is the latter we consider here. In the sentence:

The minister announced that he burned the documents an hour ago (example (24) in Smith)

Smith identifies a dependency relation of orientation between the matrix sentence (S_1) and the dependent clause (S_2) (here OT = orientation time, TE = temporal expression consisting of the combination of adverbial and tense):

"In the Orientation relation, a time in S_1 functions as OT for S_2. This means that, if one sentence is oriented to another, its RT is interpreted in relation to the other TE rather than to Speech Time. Thus in (24) the Past RT in S_2 has the relational value of anteriority, as usual; the sentence containing this RT is dependent on the matrix and the Past RT oriented to the matrix time. Because the matrix RT is future, the RT of S_2 is interpreted as anterior to the future RT. Examples like this show that dependent sentence (sic) may have an OT that is different from ST." (1981:223)

We suggest that examples like this demonstrate for *an hour ago* in:

The minister will announce at midnight that he burned the documents an hour ago

that the TA can be anchored to ST which also functions as count time or be anchored to a secondary ST located at midnight, the time at which the minister will make his announcement. In other words, the TA remains anchored to an ST, but not the ST of the sentence: in a sense the perspective of the minister (or rather the speaker reflecting his perspective) comes across in this reading (cf. Couper-Kuhlen (this volume) and Schopf (1984:Chapters 5 and 6)).

If this were not to be reflected, and the TA time is an hour before midnight, we would expect something like:

The minister will announce at midnight that he burned the documents at 11 p.m.

or:

The minister will announce at midnight that he burned the documents an hour before.

This last is in itself ambiguous, since burning could have taken place at 11 p.m. or an hour previous to some time referred to here anaphorically ('before that time we are talking about').

In simple sentences there seem to be two principle kinds of context in which TA can characterize ET without characterizing RT.

The first involves contexts in which the domain of RT does not contain ET. A number of sentences with pluperfect tense provide examples (cf. Matthews (this volume) , Schopf (this volume)). In:

Lily had spoken to Thomas at noon

the pluperfect indicates the presence of an RT anterior to ST and posterior to ET. The RT here can be viewed either more as a vantage point from which we look at ET or more as a time on the time line. Consequently, the TA time can be taken as characterizing either RT or, alternatively, ET. If the focus is on RT as a time, the TA will characterize RT, if the ET is of more interest, the TA will characterize ET. In the event that the TA time characterizes RT in:

Lily had spoken to Thomas at noon

it is already the case at noon that Lily has spoken to Thomas. If the TA time characterizes ET, at a noon anterior to RT Lily spoke to Thomas.

The second type of context in which TA can characterize ET without characterizing RT is where the TA uses RT as an anchor and count time. Thus in:

Lily had met Thomas before that

the TA refers to TA time anterior ('before') the RT referred to as "that". RT is both count time (albeit for a relation rather than a quantity) and anchor. The TA characterizes the ET as some time anterior to RT within a context-determined span. (For discussion of *before* see Heinämäki (1978:39f.), Schopf (1984:134f.), Harkness (1985:149f. and 320f.) and Hamann, forthcoming.)

In certain contexts, as in:

He won the race at five o'clock

TA time need characterize neither RT nor ET. On one reading, RT contains ET but both of these are posterior to the TA time, which is taken to provide information about which race is being referred to - the five o'clock race - rather than RT containing ET.

RT and TA time

In the next two sections we consider the interaction of different kinds of TA with RT; we divide TA into minimal interval and non-minimal interval TA for this purpose.

RT and minimal interval TA (clock times)

In simple sentences with simple past tense ET lies within or coincides with RT and both of these are anterior to ST. In the sentence:

Larry broke his leg

the event <Larry break his leg> is located at some (here unidentified) time anterior to ST, the RT can be taken as equal to ET. If we add a clock time TA:

Larry broke his leg at 8 o'clock

the TA gives us the only concrete information we have about the location of RT and it seems likely, not only on intuitive grounds, that RT is equal to the TA time. Since clock time TA are T-anchored (if not ST-anchored), the RT is to be understood as "at 8 o'clock on the day involved". TA time, therefore, = RT = ET.

However, this example involves an event notion which can be punctually conceptualized, by which we mean that English treats the event as if it is enacted at a point in time. In this and similar sentences, such as:

The train will arrive at 8 o'clock

and:

The concert begins at 8 o'clock

there is no indication that RT is greater than ET, the event can be punctually conceptualized and, in the absence of any contrary evidence, the TA time is taken as characterizing RT.

However Schopf (1984:238) notes that in sentences such as:

We'll eat dinner at 8

the ET required by the event notion is a non-minimal interval and that the TA time can only characterize the location of what Allan (1966:196f.) has called an "inceptive predication".

Clearly the intention in uttering a sentence like this is not to locate an instantaneous enactment of <we eat dinner> at the clock time. However, if the TA time characterizes RT and the event is understood as durative, ET lies beyond the bounds of RT.

If we re-interpret the event as <begin to eat dinner>, we have no such problem; the TA time characterizes the RT coincident with an ET of an event of initiation <begin to eat dinner>. The only other course open would be to take the event as durative and have TA time give something like the t_i of the RT, with most of RT and all of ET outwith TA time. There is something intuitively disturbing about a characterization which only contains a minimal part of the time being characterized. Quite apart from this, we could be forced to begin assigning t_i and t_t to times rather than events and it is not clear how this could be done.

The sentence:

I'll come and cut your hair for you at 8 o'clock

illustrates further complications. If we wish to take TA time as characterizing RT, the ET of <come and cut> would extend beyond RT. Whereas this is regularly the case with states and processes already under way (cf. Bennett/ Partee (1978:14f.) and Partee (1984:254)), this is not desirable with other event notions. We might try to focus on <come> and have TA time characterizing RT = ET of <come>, that is, in the sense of 'arrive'. This unfortunately rather neglects the business of cutting. If, on the other hand, we focus on <cut>, locating t_i of this at the clock time, we would have a situation similar to that outlined for <eat dinner at 8 o'clock> above, further complicated by the fact that <come> is not accounted for. If we interpret the events as a sequence "come and then cut", the time evidenced in the paraphrase by *then* is not covered by the TA time.

The solution sometimes adopted is to interpret <come and cut> as "come in order to cut" and then only <come> requires, at this stage, to be located on the time line via a minimal interval RT characterized by the TA. Other sentences could not be handled in this manner, however, as:

I'll cut and set your hair at 8 o'clock

demonstrates.

It is fruitless, we suggest, to reject such sentences on the grounds that the temporal structure they contain is in some sense 'sloppy'. Instead, we propose to treat <come and cut> and <cut and set> as single procedures involving more than one step. TA time in this case characterizes an RT at which ET (of an event of beginning of the procedure) is located.

In other words, durative events in combination with minimal interval TA raise the question of whether we modify the event notion, allowing us to understand the TA time as characterizing RT or take the event as durative and accept that the TA time characterizes only a small part of the RT and thus does not in the sense proposed provide a proper characterization of RT. As said, the notion of assigning a t_i to RT is questionable and, remembering the aspectual modifications of event notions with which we are familiar, we suggest that it is the event notions rather than minimal interval TA which are open to re-interpretation (cf. Schopf (1984:Chapter 4)).

This is not to deny, of course, that clock times are used with a certain vagueness. Vagueness and inexactitude are in fact the rule, since in English we seem only to order or locate events with the precision clock times afford when circumstances call for this. This is true not only of clock times proper but of a variety of expressions related to these. Were inexactitude not the rule, the familiar reassurance of *I'll be with you in a moment* or *This won't take a second/ This will only take a second* would have the opposite effect on the hearer to that intended.

Using a clock time TA to characterize a point (or more) of time in the vicinity of the TA time is, however, something very different from using a minimal interval TA to characterize what is unambiguously not a minimal interval.

Lastly, a little requires to be said about minimal interval TA in combination with states and processes. As already mentioned, if a clock time co-occurs with a state or process as, for example, in:

Lily was sleeping at 8 o'clock

and:

Bill was home at 8 o'clock

the RT is a time around and throughout which the state or process obtains. The TA time can thus be taken as characterizing a minimal interval RT (at 8 o'clock on the relevant day) around which the state and process referred to obtain.

With limited or limitable states, however, an alternative reading is possible. Since limited states can be assigned a t_i and t_t, when combined with minimal interval TA the TA can mark either t_i or t_t or, of course, indicate that at the clock time named the state obtains. Thus:

Bill was home at 8 o'clock

can be taken to indicate that at the RT of 8 o'clock <Bill be home> obtains. Alternatively, it can be taken as indicating that Bill arrived home at the time named. In this case, the TA time characterizes RT equal to the ET of an event of initiation, <begin to be at home>.

In sum, we suggest that if minimal interval TA are used to characterize RT, the RT will be approximately equal to TA time.

Non-minimal interval TA and RT

Minimal interval TA are only compatible with event notions and aspectual forms which can be tied, in some way, to a point of time. The collocational potential of non-minimal interval TA is considerably greater and, consequently, more complex.

In the sentence:

Lily stayed in London last year

the TA time is the span of the year identified as the year preceding that containing ST. If the context of utterance and preceding discourse indicate that RT is indeed the whole of this year, the TA time characterizes RT by giving us the dimensions of RT. If, on the other hand, speaker and hearer know they are talking about the Christmas break of the year previous to the year containing ST, the TA time characterizes RT by providing temporal co-ordinates within which RT lies.

Bennett/Partee (1978:22), Partee (1984:253), and Nerbonne (1983:56f.) call TA which can be used to characterize an RT of the same or less extension than TA time "frame adverbials". Nerbonne (1983:57) suggests the term reflects our intuitive grasp of a frame as "outermost". As we see it, however, their use of "frame" is more in keeping with the interval semantics approach the authors adopt. Precisely the intuitive understanding of "frame" is what is not reflected in the maximum case where the TA time is a non-minimal filled interval only suitable in contexts in which RT is equal to TA time.

Our reluctance to dub TA such as *nowadays* or *these days* "frames" is increased by the fact that a not inconsiderable number of TA involve non-minimal intervals but also carry entailments or implicatures about time outside the TA time. They might, in a sense, be thought of as casting a temporal "shadow".

These TA require us to consider not only the TA time of direct relevance to characterizing RT but also times outwith the domain of RT. *So far* and *as (of) yet*, for example, refer to time before and up to the time taken as RT but also implicate that a span of time is involved beyond TA time and thus beyond RT.

Thus:

Lily has spoken to Tom so far

can be analyzed as having ET anterior to ST, RT as vantage point at ST and the

TA time is a contextually determined span of time beginning some time anterior to ST and extending up to and including ST. The presence of *so far* indicates that in the time beyond ST (future) other events are envisaged such as, perhaps, <Lily speak to Mary> and <Lily speak to Fred>.

The T-anchored adverb *beforehand* provides a further example: TA time is a contextually determined time anterior to the anchor. In:

Lily prepared her answers beforehand

the TA is anchored to a time which is either anterior to ST or around ST and the time referred to is anterior to this again. Within the TA time lies the ET of an event. It is characteristic of events combined with *beforehand* that they are undertaken in preparation for a situation at or later than RT.

It is in connexion with the event undertaken in preparation for another event that the TA can be thought of as casting a shadow, that is, it asks us to think about times outwith or peripheral to TA time.

The shadow cast by the entailments and implicatures involved in *before, still, already,* etc. differ yet again. (For some discussion see Jessen (1974: 570ff.), Harkness (1985:206ff.), Schopf (1984:142-159), Hirtle (1977), Abraham (1980)).

Frames and intervals instead?

In contrast to Bennett/Partee, Schopf (1984:121-122) distinguishes between what he calls "interval adverbials" and "frame adverbials". Schopf's "intervals" include durative expressions of the *for X* kind (*for 3 hours*) as well as filled interval TA. His "frame adverbials" include durative expressions of the limiting kind (*within 3 hours*) as well as TA such as *during our last vacation.*

Since we consider it useful both to indicate whether a TA involves an indivisible interval or not and to reflect that otherwise non-minimal interval TA have a twofold potential, we propose to distinguish between filled interval TA (minimal or not) and other TA. An RT or ET characterized by a filled interval TA will always be equal to TA time. Other TA are able to characterize RT/ET of equal dimensions to TA time or of less extension and located within TA time.

There do not seem to be any TA in English which, if used to characterize RT, require that RT lie within the TA time and be of less extension than the

TA time. TA such as *in the middle of the night* , as we understand them, are simultaneity TA. *In the middle of the night* refers to a time described as being 'in the middle of the night', much in the way other times could be described as being 'at dawn'. Some TA, such as *meanwhile, in the meantime,* are often used in contexts where RT lies within the span of time referred to with the TA between two times (or events) given or implied. The TA themselves do not, however, require this, as is reflected by the fact that they combine readily with states.

5. TA AND PERFECT TENSES

Sufficient is said elsewhere in this volume about the details of Reichenbach's proposals and their strengths and weaknesses (see in particular Hamann) and about his and other's analyses of the perfect tenses (see in particular Matthews). Our aim is to consider the role TA play in combinations with perfect tenses and to examine whether here, too, they provide temporal settings for RT, ET, RT/ET, etc. Most of what we have to say here concerns the present perfect and the pluperfect and, since in a number of sentences we have already considered the pluperfect, the present perfect will figure more prominently. Following Reichenbach's analysis of the present perfect, RT coincides with ST:

Since we thus "know" where RT is situated and do not require a characterization to help us in this respect, we might expect TA in present perfect contexts to characterize ET. However, since there is a marked lack of collocation between present perfect tense forms and the kinds of TA ideally suited to characterize an ET anterior to RT (for example *at that time, in those days, yesterday, once*), TA in combination with the present perfect do not seem simply to characterize ET. When we then look at Reichenbach's characterization of the pluperfect in English:

we see that none of the three points of orientation coincide. RT is anterior to ST, ET anterior to RT. The pluperfect, moreover, collocates readily with both the kind of TA compatible with present perfect forms and with TA rarely or never combined with present perfect forms but regularly combined with simple past tense forms (for example, those already mentioned; *once, in those days, yesterday*).

Since the perfect tenses differ in their ability to be combined with various kinds of TA , we distinguish between them as necessary.

Kinds of TA compatible with perfect tenses

TA with a TA time consisting of a non-minimal interval either entirely anterior to its anchor or beginning anterior to the anchor and extending up to the anchor combine readily with perfect tenses; examples are *recently* and *in recent weeks*.

TA with a TA time structure which mirrors that of the present perfect, that is, where the TA time is an interval beginning anterior to the anchor (ST/RT) and extending up to the anchor, frequently combine with perfect forms. Examples are *so far, as (of) yet, hitherto,* and *until now*. We can distinguish between different kinds of these "span" interval TA on the basis of the presence or absence of implicatures about whether the span terminates at RT/ST and therefore "excludes" ST, whether the span continues at ST/RT and thus "includes" ST, or whether the TA is neutral as to whether the span terminates or continues at ST/RT (cf. Harkness (1985:162f.)).

So far and *as (of) yet* implicate inclusion of ST/RT in the span of TA time. Consequently, since the TA time is an indivisible interval, the state of affairs pertaining up to ST/RT is expected to continue beyond ST/RT.

They have been very successful so far

requires that the state of affairs <they be very successful> pertains at ST/RT.

Until-expressions, in contrast, are often understood as an indication that the span of TA time extends up to the anchor but terminates there. In contrast to:

So far the weather has been good

therefore:

Until now the weather has been good

is likely to be understood as indicating that the weather has just changed for the worst. This depends very much on the context and event notion involved however. Whereas an *until*-expression and simple past clearly indicate that the TA span extends up to the anchor but terminates there, as in:

Until now I supported Norbert for the job

with the inclusion the present perfect offers, TA time can also include ST/RT. We could therefore continue:

I have supported Norbert until now and don't intend to change.

What is required for all of these span TA with present perfect tense, however, is that the state of affairs pertaining over the indivisible interval be under discussion at ST/RT.

A number of these TA carry no implicatures about whether the span terminates at the anchor or not. Into this category fall TA such as *hitherto/thitherto, up to now/up to then, in the past*. As discussed elsewhere (Harkness (1985), Matthews (this volume)), distinctions can be noted between these TA in terms of the kind of event notion likely to be combined with a given TA span and whether one event holds over the span or a series of repetitions of one event. *Hitherto/thitherto*, for example, require that the span of time hold for a series of instances of a given event. Thus:

We have been very lucky hitherto

in contrast to:

?We have been very happy hitherto

allows more readily a reading "on every occasion where luck/misfortune were possible".

The compatibility of *since*-expressions with present perfect and pluperfect forms is well documented. *Since*-expressions differ from the filled interval TA just considered in two major respects. First, the time at which the span extending up to the anchor begins is named directly or via an event in the *since*-expressions. *Since Christmas* names some Christmas (which the context specifies) as the time at which the span begins: *since they left* refers to an event <they leave>, itself locatable on the time line.

The second major difference is that the TA time is not a filled interval, it can therefore characterize an ET of the same extension as TA time or of less extension and located within TA time. It is often not possible to tell without further information which reading is most appropriate in a given context.

Since-expressions are most commonly found with perfect forms (if *since* introduces a temporal clause, with perfect in the main clause). Modal constructions, however, can take present tense forms with *since*-expressions. Thus we can have both:

Since last night he cannot move his fingers properly

and:

Since last night he has not been able to move his fingers properly

or:

He won't eat since Saturday

and:

He's refused to eat since Saturday.

Dowty (1979:344) discusses the idea that time adverbials with the present perfect characterize an 'extended now'.[9] Whereas we would prefer to say that TA characterize the span of time up to ST referred to with the present perfect (and thus have ET clearly anterior to 'present time') we do think that the combination of *since* TA with present tense forms comes very close to characterizing an 'extended present'. Consequently, <he will-not eat> above can be seen as the 'extended present' outcome or product of an event (or events) of refusing to eat anterior to ST/RT. To put it another way, the state of

9 He takes up a proposal made among others by McCoard (1978) that the present perfect can refer to an "extended now".

affairs over the 'extended present' is initiated by some refusal to eat.

This could go some way towards explaining the causal connexion between the state of affairs at ST/RT and the time named in the TA (or more probably, events at this time). Consider, for example, the following pairs of sentences:

Since lunchtime Lily is all smiles
Since lunchtime Lily has been all smiles

Since then he drives faster
Since then he has driven faster

and, to take an example from Leech (1971:45):

I am walking to work since my car broke down
I have been walking to work since my car broke down.

In each case the sentence with present perfect can be read as simply providing the span of time up to ST/RT over which a certain event holds. Although in our examples we tried to avoid the likely causal connexion of Leech's example, we note that in all the present tense sentences the implication is that the state of affairs at ST/RT is in some way the outcome or result of events at the time named in the *since*-expression. This would also explain the preference for initial position for the TA in these sentences; in the present perfect tense sentences, the TA could equally well have taken final position.

Present time TA

The TA we have considered so far provide a setting for the ET anterior to ST/RT by referring to a span which extends up to ST/RT. Now we wish to consider TA which characterize RT, that is, which refer to the time around ST/RT. Consider in this light the following:

We have done all we can { for the moment / right now / for the time being / now }

In these examples, we suggest, ET is anterior to ST/RT (at a time not characterized here) and the TA time refers to ST/RT. Whether all of the TA above are ST-anchored or merely T-anchored is of no consequence here, they refer here to an extended present which includes ST/RT. They stand in contrast therefore to one reading of:

We have done all we can today

in which the TA time provides a setting for the ET rather than a characterization of ST/RT.

For the moment, for the time being and *right now* (meaning 'for the present') are filled intervals and the state of affairs pertaining over the filled interval is the outcome, consequence or resultant state of the foregoing event(s). In the examples above <we do all we can> precedes the present state of affairs in which all that can be done within the present time has already been done (cf. Matthews (this volume)). In this respect they resemble *since* TA plus present tense.

Other TA which include ST

TA such as *this week, today, this year* can be used to refer to time posterior, anterior or coincident (overlapping) with ST:

I was/have been in Paris this year
I am working in Paris this year
I will be in Paris this year.

Unlike the simple past in:

Arnim cut his finger today

only durative event notions which hold over the time span indicated by the interaction of TA time, present perfect and the context of utterance can occur with present perfect forms. Consequently, the rather unusual:

Arnim has cut his finger today

would have to be understood as some sort of accomplishment, the resultant state of which holds over the relevant interval. This is then rather like:

I have run a mile in three minutes today

or the accomplishment interpretation of what is normally an achievement in:

I have beaten Roger (at tennis) today.

Limitable states and processes extending exclusively or inclusively up to ST/RT combine with these TA and present perfect forms:

I have been very busy today
Irene has been in London all week.

Filled interval TA including ST do not combine well with the present perfect:

* I have been very busy these days
* People have been leaving school later nowadays.

TA times which exclude ST/RT

We have suggested that only TA compatible with a Reichenbachian analysis of the present perfect are compatible with the present perfect. However, present perfect forms are found with TA which refer to time entirely discrete from RT:

He has been there in September
I've tried it out at five o'clock.

In these examples the TA does not characterize time extending up to RT within which ET must be posited. Instead, the TA should be understood as providing more information about the temporal profile of the event under discussion.

TA which are incompatible with the Reichenbachian configuration are therefore compatible with present perfect forms if the TA characterizes something other than the span containing ET and extending up to RT. In such contexts, these TA are without the anchor they require to be able to locate events in time.

TA incompatible with the present perfect

TA with a TA time which is entirely anterior to ST/RT are only compatible with present perfect tense forms if not used to characterize ET. Thus *once, yesterday, in 1919* are not readily combined with the present perfect. TA which use a time distinct from ST/RT as anchor are also not readily combined with present perfect. Examples of such TA, often found with pluperfect forms, are *previously, before that*.

If, however, the present perfect is part of a narrative present sequence, many of these are possible. The following constructed text should illustrate this point:

Let us imagine you are one of Portia's suitors. The time is approaching when you have to choose one of the caskets. *Previous to* setting out, you *have prepared yourself* as well as possible. You *have* even *once consulted* the court astrologer and *before being introduced* to Portia you *have studied* her tastes carefully.

Perfect or simple past?

A number of non-minimal interval TA occur with both simple past and present perfect tense forms and, therefore, with pluperfect forms. A well-known example is *recently*. *Of late*, in contrast, involves a TA time which necessarily extends up to and includes RT and does not therefore occur with simple past tense forms. For some TA, however, occurrence with simple past or perfect forms seems to differ from one variety of English to another. Thus Quirk et al (1972:488) suggest that *lately* occurs with both simple past forms (and hence can characterize an RT/ET entirely anterior to ST) and present perfect forms. For British English speakers *lately* involves an indivisible interval which extends up to ST/RT and it occurs only with the present perfect. Quirk et al suggest American English speakers also use it with simple past and provide the example:

He lately moved into a new appartment.

Jacobson (1964:288) has an example from his corpus in which *lately* occurs in a pluperfect context:

Connie, an ARP warden, had lately moved into one of the two top floors of the Chilcome Street house.

The resemblance between this and the example in Quirk et al is striking. As mentioned above, TA which do not occur either with present perfect or simple past tense forms do, nevertheless, occur with pluperfect. The combination with pluperfect tense in the sentence above is perfectly possible for British speakers. Unfortunately, any American informants we have had have not agreed with Quirk et al (cf. Dowty (1979:341)). On the other hand, the minimizing adverb *just* exhibits precisely this pattern of distribution (cf.

Matthews (this volume)). British speakers tend to use it with present perfect forms, that is to locate an ET somewhere within a span of time proximal to ST/RT:

I have just seen Lily

whereas American speakers regularly combine *just* with the simple past:

I just saw Lily.

TA and the pluperfect

Since pluperfect tense forms are the equivalent of a back-shifted simple past but are also the forms required to indicate in a marked fashion that one of two events both anterior to ST is anterior to the other event, they are compatible with types of TA which prefer or require the present perfect rather than simple past as well as with those which occur with simple past.

ST-anchored TA such as *until now, hitherto, yesterday, today,* require the appropriate narrative mode to occur with the pluperfect. Their non-ST-anchored equivalents are regularly found with pluperfect forms since they mirror the Reichenbachian interpretation of the pluperfect: *until then, thitherto, the day before (that), that day.* Whereas non-ST-anchored TA occur readily with pluperfect forms, the majority can also be found with simple past: *the day before, three days before.* However, *the previous day* and *three day previously,* which do not have the possibility of anaphoric reference to an anchor open to *before*-expressions, do seem to require either pluperfect forms taken as indicating that they are not anchored to ST, or, with simple past tense forms, an anchor which is not ST.

This is how we understand the anchoring of *previously* in the following example collected by Jacobson (1964:320):

The child finds himself in some situation which elicits recall of something which was previously a part of that situation but which is now absent.

Summing up the interpretation of TA with pluperfect forms, we can say that the TA have three major functions: either TA time characterizes RT excluding ET (parallel to Reichenbach's interpretation of the pluperfect) or the TA can characterize ET entirely anterior to RT (with the TA often anchored to

RT) or, as we saw for present perfect tense forms, the TA can characterize a span of time beginning anterior to RT and extending up to RT. Within this span ET is to be located. In the last case, ET can either be equal to the TA time or contained within it, as alternative readings of:

Lily had lost weight since Christmas

indicate.

We saw too that certain TA use RT as an anchor and count time (*the week before that, the previous week*). In:

Lily had lost weight { the previous week
{ the week before (that)

the most salient reading is perhaps that the TA time is taken as the week anterior to the week containing RT and <Lily lose weight> is located within or over TA time. We could, however, go one step further and view the TA as anchored to an RT and providing us with time one week anterior. At a time within this week it was already the case that Lily had lost weight. In German this reading is the only possible for:

Die Woche zuvor hatte Lily Gewicht verloren gehabt.

This brings us to Reichenbach's comments on *yesterday* in the sentence:

I had met him yesterday

"What was yesterday is the reference point and the meeting may have occurred the day before yesterday". (1947:294) We can now see that this is only one possible interpretation. RT could be some time provided by the context and ET could be located within the TA time of a day one day previous to the day containing a secondary ST/RT.

6. TA AND DIRECT AND INDIRECT REFERENCE

It will be clear from the foregoing that we do not require TA in order to refer to definite times. Very often the presence of a TA simply tells us about the temporal setting of the times referred to, sometimes the presence of a TA will indicate that an indefinite reading is not called for (as, for example, with the present perfect when either reading would otherwise be possible).

There are two TA of which we are aware which signal that the temporal reference is indefinite, these are *one day*, as in:

Lily could be famous one day , you know

and *sometime*, as in:

Come up and see me sometime, won't you?

In view of the roles played by TA in temporal reference, it is hardly surprising that so few TA indicate indefinite reference. Frequency expressions, such as *sometimes, (up)on occasion, now and then*, all of which entail a frequency score of more than once, regularly indicate that indefinite temporal reference is involved.

Since we require of TA that they provide settings, the definite times referred to can be equal to TA time but also simply located within TA time. In a sentence like:

It rained during the night

RT/ET is located within the setting of TA time. We can imagine the speaker was awake when it rained and hence knows precisely what time he/she is referring to. However, in:

It must have rained during the night

'must have' indicates that the speaker does not know at what time it rained nor, presumably, whether it rained several times during the night. Reference here is, we suggest, nevertheless definite. The speaker is not referring to any time but to whatever time within TA time is RT/ET. The fact that the speaker cannot locate <it rain> on the time line beyond that it is within TA time and does not know whether it rained once or several times does not mean that the temporal reference is "indefinite" as we understand it. What is being said is that there is a time within TA time and for this time <it rain> holds.

Leech (1971:32) points out that a number of expressions combine with the present perfect when indefinite temporal reference is involved. He mentions in this context the quantifiers *ever* and *never* as well as *before (now)*. It is certainly correct that sentences such as:

Has Lily ever been to the Louvre?

and:

Lily has never been to the Louvre

do not refer to definite times. *Before(now)* differs somewhat from these non-TA expressions. In:

Lily has been to the Louvre before(now)

no reference is made to a definite time anterior to ST/RT at which Lily visited the Louvre. We should note, however, that the sentence requires that some definite visit to the Louvre be under discussion at ST/RT (cf. Harkness (1985: 151f.)).

Concluding remarks

In setting out to examine the role TA play in temporal reference, we have restricted ourselves to the simple sentence. Of interest now would be to consider the ways in which TA are employed in extended structures - in complex sentences and across sentences - when more than one RT and ST are often involved.

The emphasis given here to the role TA play has left little opportunity to discuss the temporal structure of individual TA in any detail. Precisely this, however, explains the appearance of one TA rather than another in a given context. Synonymity among TA is as rare as elsewhere in the lexicon (cf. Harkness (1985) and forthcoming).

PRESENT PERFECT TENSES: TOWARDS AN INTEGRATED FUNCTIONAL ACCOUNT

Richard Matthews

0. Introduction

There is nothing so elusive as grammatical meaning, and the meaning of the present perfect tenses is no exception. However, some attempt to arrive at it needs to be made if we are to be able to examine the effects of collocation with temporal adverbials or to explain the non-occurrence of certain classes of predicates in English Present Perfect (PresPerf) and Present Perfect Progressive (PresPerfProg).

This paper[1] has developed out of difficulties I encountered in establishing readings of non-finite perfect forms, viz. Perfect Infinitive and Perfect Progressive Infinitive as complements of Modal Auxiliaries. If one wishes to ascertain whether modal and non-modal forms differ semantically in ways that are not accountable for entirely in terms of the modal vs. non-modal distinction, then one has to have a clear idea of what the semantic function of perfect forms is.[2] This is not exactly straightforward when the standard accounts in

---

1 The original version of this was presented in a project meeting on 18 July, 1985. In revising it for print, which has involved a great deal of reworking, I have referred to a wider range of accounts of the Perfect and also taken account of the new presentation of tense and aspect in Quirk et al. 1985, which I had not had an opportunity of consulting earlier.

2 The well-attested neutralization of the co-occurrence restriction with Perfect and Past tense forms in modalized predications:

John has helped Mary	already/before now
	*yesterday/*in those days
John helped Mary (AmE)	*already/*before now
	yesterday/in those days
John may have helped Mary	already/before now
	yesterday/in those days

suggests that both 'indefinite past' readings (associated with finite Perfects) and 'definite past' readings (associated with Past tenses) can be ascribed to non-finite Perfects. This follows, however, only if we adhere to the polysemic view discussed in 1, and only if we take the extreme view that tense-forms are only interpretable in terms of collocating adverbials, whether present in immediate context or not.

descriptive linguistics are by no means uniform: though they may have a number of features in common, they are not only occasionally contradictory, worse still, they rarely employ any degree of methodological rigour in describing the semantics of these forms.

In the following, therefore, I wish to examine some of the claims made about perfect forms and then attempt to produce an integrated account that is linguistically sounder as well as a viable basis for the examination of the interaction between perfect tense forms and adverbials, and also (though not in this paper) of the interaction between modality and the temporal reference of non-finite complements.

1.0 Monosemic and polysemic views of perfect tenses

If we examine the claims that have been made about the Present Perfect over roughly the last century, two basic positions seem to emerge, irrespective of the details of the description.

1. the PresPerf has one basic meaning (which may be modified by certain pragmatic conditions) which interrelates with the lexical meaning of the predicate, i.e.

   meaning of PresPerf $\cap$ meaning of predicate
   = meaning of verbal group
   This we may term 'the monosemic view'

2. the PresPerf has a number of meanings which (though perhaps historically related) are discrete and contrastable. Certain predicates (or predicate classes) are restricted as to which meanings they are compatible with, i.e.

   predicate as PresPerf^1, PresPerf^2, etc.
   This we will call 'the polysemic view'

The Present Perfect Progressive has, admittedly, received much less attention, but here, too, we can establish a basic dichotomy of opinion.

1. PresPerfProg differs from PresPerf in a way that is predictable from and parallel to the difference in meaning between progressive and non-progressive forms in general. This may be referred to as 'the aspectual modification view'

2. PresPerfProg has a meaning (or a set of meanings) parallel neither to PresPerf nor to other progressive forms. This may be referred to as 'the autonomous tense view'

It should be pointed out that some writers make a distinction between the 'meaning' of a tense and the 'use' or 'uses' of a tense, thus making it possible to accept in principle the monosemic view while describing and analyzing a number of uses of PresPerf. We should be a little chary of this use of 'use'. In morpho-syntactic terms, it makes perfect sense to say, e.g. the PresPerf is used in syntagm X, but to say, e.g. the PresPerf is used to express 'Y', is clearly a statement about the semantics of the PresPerf, and it would be better to employ 'sub-sense' or even 'sense', if theoretical claims about PresPerf are, in fact, being made. Huddleston, who also avails himself of this usage, insists that 'uses' are "pre-theoretical": observations about the semantics of a form, not the description of the semantics. In the section that follows I shall employ the term 'reading' rather than 'sense', 'meaning', 'basic meaning' or 'use'.

We can find claims for a one-, two-, three-, or even four-reading view of PresPerf. Without entering upon a major survey, I will cite a selection of those representing certain positions and, where appropriate, indicate definitions. Of recent accounts, there is a certain orthodoxy about the three- or four-reading PresPerf: earlier accounts tend to have favoured one or two readings.

1.1 The Present Perfect

1.1.1 Four-reading accounts

This kind of account we will have reason to return to in a later section (1.3). I will outline here accounts to be found in: McCawley 1971, Comrie 1976, Harris 1984, Huddleston 1984 and Leech 1969 and 1971.

Leech 1969 distinguishes "four separate senses" (Leech 1969:153) for each of which a formal specification is given rather than a definition. Listed below are Leech's descriptions:

L 1. Duration of state up to the present moment
2. Duration of habit up to the present moment
3. Indefinite Past [3]
4. Resultative Perfective

The same differentiation also appears, with slightly different nomenclature, in the more practically oriented Leech 1971:

L 1'. State-up-to-present
2'. Habit-in-a-period-leading-up-to-the-present
3'. Indefinite Past
4'. Resultative Past

3 Leech actually gives two formulae for this, one of which could be said to define a 'recent past'. One specification includes the other, or may be inferred from the other.

McCawley's four-reading account (McCawley 1971) is not isomorphic with Leech's:

McC	1.	Universal
	2.	Hot News
	3.	Existential
	4.	Stative

McCawley's 'universal' covers Leech's readings 1 and 2, Leech's 'indefinite past' covers McCawley's 'hot news' and 'existential'. Comrie's 1976 and Harris's 1984 discussions arrive at essentially the same set but use different names.

C	1.	Perfect of persistent situation
	2.	Perfect of recent past
	3.	Experiential perfect
	4.	Perfect of result

H	1.	Extended now time	(= C 1)
	2.	Hot news	(= C 2)
	3.	Indefinite anterior	(= C 3)
	4.	Resultative	(= C 4)

Another variant of the four reading approach is to be found in Palmer/Blandford 1939. They distinguish:

P/B
1. "an action completed at the moment of speaking, but not associated with any adverbial (expressed or understood) of past time"
2. "an action still continuing at the moment of speaking, but associated with an adverbial (expressed or understood) of anterior duration"
3. "general anteriority"
4. "repeated actions associated with past time"

Of P/B 2 we may note that it is claimed that if a verb is progressivezable (and the sentence is affirmative) then PresPerfProg will occur.

Finally, Huddleston 1984 distinguishes four "uses" corresponding to the McCawley/Comrie/Harris differentiation.

Hd
1. "state of affairs prevail(ing) continuously throughout a period beginning in the past and extending up to the present"
2. "recent events"
3. "past events where the actual time of occurrence is unimportant"
4. "past events considered as causes of their present effects or results"

(Huddleston 1984:160-161)

Huddleston, however, is committed to the view that these uses "merely illustrate some of the ways a past situation may be connected with the present so as to motivate our representing it as located in an inclusive rather than an exclusive past" (Huddleston 1984:161).

1.1.2 Three-reading accounts

Examples of these are to be found in Schmole 1977/8, Zandvoort/van Ek 1975, and Quirk et al. 1985. Schmole and Zandvoort/van Ek are more or less equivalent, the only difference being that Schmole explicitly limits his 'resultative perfect' to cases where results are entailed as holding at the moment of speaking.

Z	1.	continuative perfect	=	Sch. 1.	continuative
	2.	resultative perfect		2.	resultative
	3.	perfect of experience		3.	perfect of experience/ indefinite past

Quirk et al. 1985 presents what is essentially a revision of the Leech/ Svartvik 1975 account of tense meaning. This allows for a "common factor of meaning" (the "anterior time zone") for "perfective (sic) constructions" (Quirk et al. 1985:192), while discriminating three meanings:

Q 1. "state leading up to the present"
2. "habit in a period leading up to the present"
3. "indefinite event(s) in a period leading up to the present"

(Quirk et al. 1985:193 - the order has been changed to facilitate comparison with other accounts). The event meaning is linked to a set of three implications:

a. "relevant time zone leads up to the present"
b. "the event is recent"
c. "the result of the action still obtains in the present" (ibid.)

We may infer from this that Quirk et al's 1 and 2 represent continuative meanings, while their reading 3 includes both resultative and indefinite/ experiential readings.

1.1.3 Two-reading accounts

These are to be found particularly in earlier accounts, e.g. Sweet 1900, Jespersen 1924, 1933, but more recently also in Palmer/Blandford (revised Kingdon) 1969, and Close 1975. The differentiation in Sweet and Jespersen is essentially the same.

S	1.	incomplete perfect	=	J 1.	inclusive present
	2.	complete perfect		2.	permansive present ("present state as outcome of past events")

Similar, though explicitly linked to a distinction between states and actions, are the definitions in Palmer/Blandford/Kingdon and Close.

P/B/K 1. "state or condition that still exists began ... in the past"
2. "action performed (and ... completed) at some unspecified time in the past"

Cl 1. "activity continuing until now"
2. "activity in a period continuing until now"

We may identify the first reading in each of the above cases with a 'continuative perfect' and understand the second as covering both 'resultative' and 'experiential' in the 3- and 4-reading views.

1.1.4 One-reading accounts

The one-reading view is upheld by Twaddell 1963, Sørensen 1964, Joos 1968, Bauer 1970, Onions 1971, F. R. Palmer 1974, McCoard 1978 inter alia. In addition, a number of other scholars have accepted the "basic unity" of the Perfect, while still delimiting a number of readings (e.g. Huddleston 1984, Quirk et al. 1985). We may divide adherents of the one-reading view according to the kind of definition suggested.

Onions 1971 defines:

"The *perfect*, *pluperfect*, and *future perfect* are used to present, past or future state (respectively) resulting from an existing or continued action regarded as completed at the time of utterance"
(Onions 1971:103)

A similar emphasis on completion appears in Huddleston 1984:

"'Perfect' is to be understood in the sense "complete": in its strictest use, perfect will thus be applied to a formal category of the verb or VP whose primary meaning is to denote the state characterized by the completion of some earlier situation"
(Huddleston 1984:163)

Huddleston admits, however, that certain uses of PresPerf cannot be interpreted as involving completion (e.g. continuative readings). To cover this, Huddleston, like McCoard 1978, avails himself of the notion 'inclusive past', i.e. past inclusive of the present (cf. Jespersen's notion of 'inclusive time' in Jespersen 1924:271-2).

Palmer 1974 and Twaddell 1963 select 'current relevance':

"'current relevance' - that in some way or other (not necessarily in its results) the action is relevant to something observable at the present"
(Palmer 1974:50)

cf. Comrie's general meaning: "continuing relevance of a previous situation" (Comrie 1976:56).

> "Modification II, have + participle, explicitly links an earlier event or state with the current situation. It signals a significant persistence of results, a contained truth value, a valid present relevance of the effects of earlier events, the continued reliability of conclusions based on earlier behavior.
> ... Put negatively, have + participle asserts that there has not been any intervening change to affect importantly the validity (or the inferences from the report) of an earlier event or condition ...
> The have + participle modification does not necessarily imply completed action except to the extent that the earlier 'action' must have matured sufficiently to produce currently relevant effects. There need be no implication of completion in the sense of non-continuation of the 'action' at the present moment. The term 'perfect' has been unhelpful by suggesting some kind of completedness, previous termination, as essential semantic ingredients of Modification II, whereas these meanings are at most compatible by-products of the semantic combination of context, situation, the lexical verb, and the 'Current relevance' signal of have + participle."
>
> (Twaddell 1963:8-9)

We may also compare Joos 1968:140:

> "the meaning of the perfect phase (sic) is that the principal effects of the event are out of phase with it, which of course can only be true if they are delayed"

McCoard 1978 makes an extensive survey of views of the Perfect and classifies theories in terms of four categories:

a. current relevance (CR)
b. indefinite past (ID)
c. extended now (XN)
d. embedded past (EB)

These are characterized:

a. "expresses a present state resulting from past action"
b. "expresses a past event which is unidentified as to time"
c. "expresses a past event within a time span which is continuous with the present, not differentiated into "then" versus "now""
d. "is made up of a past-tense sentence embedded as sentential subject of a present-tense sentence"

(McCoard 1978:18)

This classification is at best somewhat question-begging, at worst misleading. 'Current relevance' is characterized as a resultative perfect; the 'embedded past' is characterized in terms of a TG model - in fact non-TG classifications such as Halliday's 'Past-in-Present' should perhaps be included under this (Halliday 1976:153). Moreover, it seems that 'EB' is really a variant of 'XN'.

McCoard argues that 'XN' "offers the best overall account of the perfect/preterit opposition and attendant phenomena" (McCoard 1978:18) while admitting (p. 204) "... we find that EB theory can be interpreted - with some difficulty - as a version of XN theory". (It should be noted that McCoard's 'extended now' is not equatable with Harris's 'extended now' which is a synonym for 'continuative'.) In point of fact, better definitions of McCoard's view of the Perfect can be found.

> "the perfect locates the situation within a period of time beginning in the past and extending forward to include the present (TU (= time of utterance), or more generally, TR (= time to which the time of a situation is related)"
>
> (Huddleston 1984:158)

This is reminiscent of Reichenbach 1947, which I shall consider at a later stage.

Quirk et al. 1985, in identifying an "anterior time zone" (cf. above) as basic to the Perfect are adopting a view of tense (though they call it 'aspect') also to be found in R. L. Allen 1966, which is compatible with the above view.

Sørensen 1964 should be singled out as one of the most categorical (and argumentationally convincing) upholders of the monosemic view. I shall return to some of his arguments in due course, but we may note here his definition of PresPerf, which is also influenced by Reichenbach 1947:

> "[It] denotes that the action denoted by the stem of the (main) verb is an accomplished fact at the moment of speech, and, consequently, that the action is anterior to the moment of speech. (Note this: The perfect does not denote that the action is brought to a conclusion at the moment of speech. Nor does it denote that it is not brought to a conclusion. This is something which has nothing to do with the perfect qua perfect ...)"
>
> (ibid: 74)

> "... a two-point tense - a point-of-action tense, such that the point of the action is unspecifiable, since the perfect is incombinable with an indication of past time, and a point-of-reference tense, such that the point of reference is the moment of speech, whereas the point of the action is anterior to and never includes the moment of speech ..."
>
> (ibid: 83)

> "There may be cases which require special treatment, but essentially the perfect is a very simple tense in English, a tense with one and only one notional value: 'A has done such and such a thing ...' means 'A's doing such and such a thing ... is now an accomplished fact', and that is all it means."
>
> (ibid: 82-3)

Bauer 1970 accepts that Sørensen is basically correct, and defines PresPerf in very similar terms:

"the action is viewed, not as a past event, but as being an accomplished fact at the moment of speaking, having taken place, once or repeatedly, within a span of time that is not conceived as separated from the moment of speaking."

(Bauer 1970:190)

"the action, anterior as such to a certain point of reference (frequently, but not necessarily, the moment of speaking) is viewed as being, at this point of reference, an accomplished fact."

(ibid: 191)

Bauer adheres to the term 'resultative', however, acknowledging that results may be only implicit or even (as F. R. Palmer says) 'nil results'. This is for him the 'normal case' or 'principal variant', while the 'continuative' perfect is a 'special case' or 'combinatory variant'. The 'perfect of experience' is "to be regarded as a type of context (the relating of a person's experience), but not as a special function of the grammatical category perfect" (Bauer 1970:191).

What Bauer adds to Sørensen's argument is the discovery that the telic vs. atelic distinction (following Garey 1957) can be taken as an indicator of how a PresPerf predicate will be interpreted when collocated with an adverbial such as *up till now*: 'telics' will give iterative readings, 'atelics' continuative readings. Bauer includes 'momentaneous actions' (cf. Schopf's punctual events) under 'telic'. Jespersen's terms 'conclusive' and 'non-conclusive' (Jespersen 1924:273) may be better. These are the terms adopted in Quirk et al. 1985:201. It will be seen that this claim is essentially confirmed by the tests discussed in 2.1.1, though the situation is a little more complex.

1.1.5 Summary

We have attempted to give an overview of accounts of PresPerf in Table I (cf. p.120). The relationships between the various readings look very much as if some kind of semantic hierarchy obtains, but we will not go so far as to claim that this is the case until examination of test predications has been undertaken (cf. 2).

1.2 The Present Perfect Progressive

In comparison to pronouncements on PresPerf, there are very few on PresPerfProg. Those who view PresPerfProg as simply an aspectual modification of PresPerf parallel to all other progressives include Sweet 1900, who terms the meaning

Table I: PresPerf readings - an overview

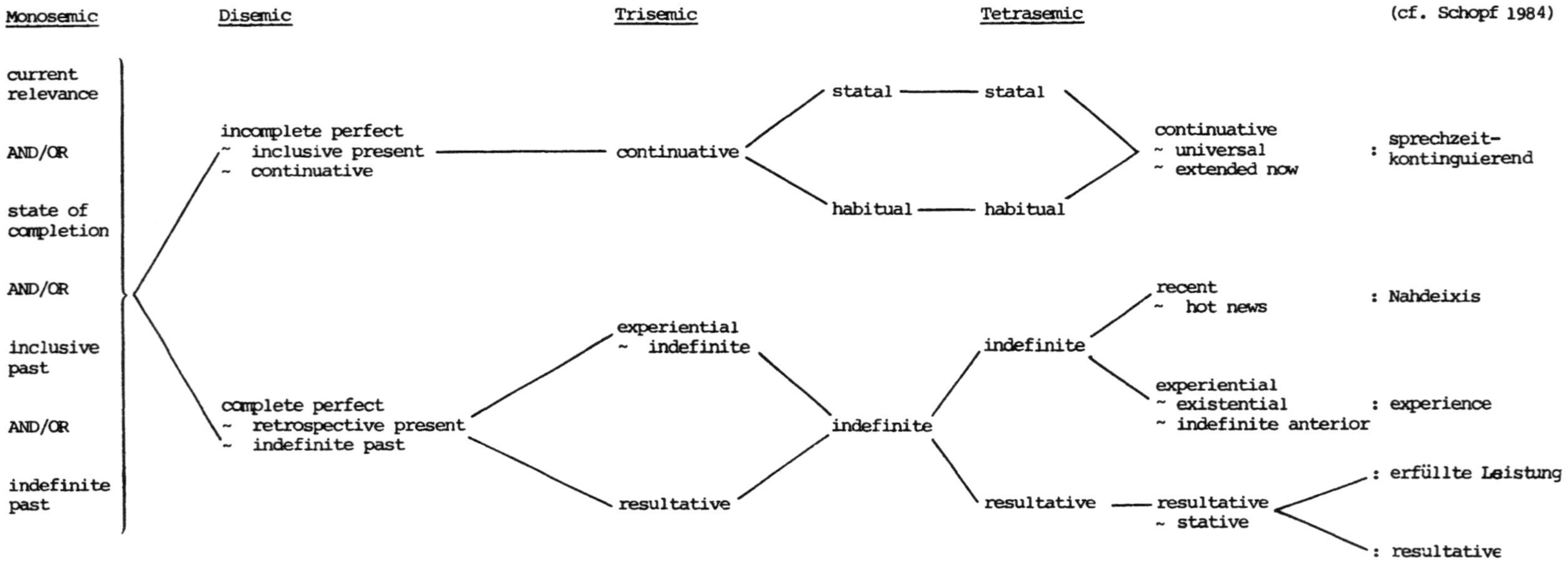

of progressives 'definite', without, however, giving any kind of coherent or plausible definition of the term, and F. R. Palmer 1974, who says:

> "The position (of perfect progressives) is exactly the same as with the non-progressive forms, provided only that it is accepted that the use of the progressive does not necessarily imply that the activity continues *throughout* the relevant period of time, but merely that it has duration *within* the period."
>
> (F. R. Palmer 1974:53-54)

Bauer 1970 is also among those who see PresPerfProg as an aspectual modification of PresPerf. What it expresses is "simply 'accomplished fact that action has been in progress at, or for, some time within the time-sphere of the perfect, the span of time immediately anterior to the point of reference'" (Bauer 1970:195). 'Telics' become 'quasi-atelic', i.e. imperfective, while with 'atelics', progressive aspect may be "felt to be redundant" (ibid:195) or give rise to "very subtle (semantic) distinction" (ibid:196).

On the other hand, we have accounts such as Leech 1969, where PresPerfProg is associated with only 2 of his four readings of PresPerf:

1. duration of state up to the present moment
2. duration of habit up to the present moment

(Leech 1969:154)

and that of Schmole 1977/8, who handles the progressive vs. non-progressive aspectual distinction in terms of meaning components, rather than holistically, at least initially, and then suggests which PresPerf senses do not have correlates with PresPerfProg. Thus, for Schmole, resultative and experiential perfect do not work with progressive aspect; with habitual continuative perfects, PresPerfProg and PresPerf are interchangeable, while non-habitual continuative perfects leave room for "neue Bedeutungsdifferenzierung" - neither of these involve actions, events or states-of-affairs that necessarily include the moment of speaking. Where the moment of speaking is included in the time of the action ("Handlungen, die als in R_1 einmündend markiert sind") the PresPerfProg is obligatory (Schmole 1977:254). Finally, Schmole differentiates cases like *I've been skiing* as a subsidiary function of PresPerfProg, though it might be objected that these are PresPerf of the lexically complex: *go skiing* (cf. German *skifahren, ich war skifahren,* etc.) If we exclude this, we are left, effectively, with 3 readings:

1. habit continuing up to the present
2. activity continuing up to the present
3. recent activity with "concomitant overlapping"

'Concomitant overlapping' (R. L. Allen's 1962 term) means no more and no less than certain effects of the activity are still perceivable at the moment of speaking. This is very close to what some scholars include under 'resultative' with PresPerf, but Schmole's restriction of 'resultative' to entailed results prohibits identification of the two. Apart from this, Schmole's account is similar to Leech/Svartvik 1975:

1. temporary state up to the present
2. temporary habit up to the present
3. temporary (activity) with present results

(author's adaptation)

Quirk et al. 1985 establish two readings:

1. temporary situation leading up to the present
2. temporary habit up to the present

and a number of implications, which they formulate, informally, as:

a. (limited) duration
b. continues up to the present or recent past
c. need not be complete
d. may have effects which are still apparent

They claim that PresPerfProg combines the meanings of the perfect with those of the progressive (i.e. duration, limitation of duration, possible completeness), but note:

> "Nevertheless, the perfective (sic) progressive has a semantic range that is not entirely predictable from the meaning of its components."
> (Quirk et al. 1985:210-211)

Implications of PresPerfProg also figure in Onions 1971, who appears to adopt a monosemic view ("action as going on/state as existing" and thus, implicitly "non-completed"). It is noted that PresPerfProg often has an "emotional colouring" of surprise, disgust, etc. as in *Someone has been tampering with the lock*. Close (1975:250) suggests that PresPerfProg denotes an "activity in progress in pre-present period" which may either be concluded or continue up to the moment of speaking. It has implications of temporariness and-or incompleteness with the possibility of concomitant overlapping.

Finally, Comrie, who appears to adopt a tetrasemic view of PresPerf, relates at least three of his readings with PresPerfProg, viz.

1. persistant situation
2. experential perfect [4]
3. recent past

4 This seems to work only with *ever* and *never*.

He makes no mention of a progressive form with a resultative meaning (concomitant overlapping), but since his discussion of the perfect of result covers 'perfecto-praesentia' like *He is gone, He's got a house*, which suggests a restriction like Schmole's on the 'resultative perfect' (entailed result), it is to be supposed that cases of 'concomitant overlapping' would be included under 'recent past'.

It is less easy to produce a comparative display for PresPerfProg, as we did for PresPerf in Table I, but nevertheless an attempt has been made in Table II (cf. p.124).

1.3 Support for the tetrasemic view?

Support for the tetrasemic view of PresPerf comes from cross-linguistic comparison. It is a frequently recorded fact, for example, that a 'continuative' perfect typically corresponds to present non-perfect tenses in French and German. *I've been waiting for an hour ≃ j'attends depuis une heure ≃ ich warte seit einer Stunde*. More compelling is the nature of equivalence between varieties of English, especially when some measure of diglossia is common. Harris 1984 compares the Standard English PresPerf with its four equivalents in Irish English (Hiberno-English): Past (alias Preterite); BE + *after* + V-*ing*; transitive HAVE + V-en/ intransitive BE + V-en; and Present. These, he claims, are traceable back to Irish Gaelic and Elizabethan English. I have summarized his comparisons, both synchronic and diachronic, in Table III (cf. p.125).

One weakness in Harris's comparison is that he does not include the Standard English resultant-stative constructions: HAVE + N + V-*en* (favoured more in American English) and HAVE *got* + N + V-*en* (favoured more in British English) - their existence, however, is mentioned.

The corroboration of the tetrasemic view of the perfect is only apparent, however. Without anticipating the arguments of the following section, we may suggest two reasons why the cross-dialect comparison cannot help us to resolve the monosemic vs. polysemic question. One is that the kind of classification Harris, inter alia, employs fails to distinguish grammatical meaning (as a part of semantics) from utterance meaning. Thus 'hot news' is an apt term not for the meaning of the Perfect, but for the nature of the message it is contained in. Casting the assertion of the occurrence of an event in the Perfect does not ipso facto make it 'hot news': it is 'news' if it was not known to the addressee before, and it is 'hot' if the event is recent, but that is not the function of the Perfect. Furthermore, unless we adopt the entailed result

Table II: PresPerfProg readings - an overview

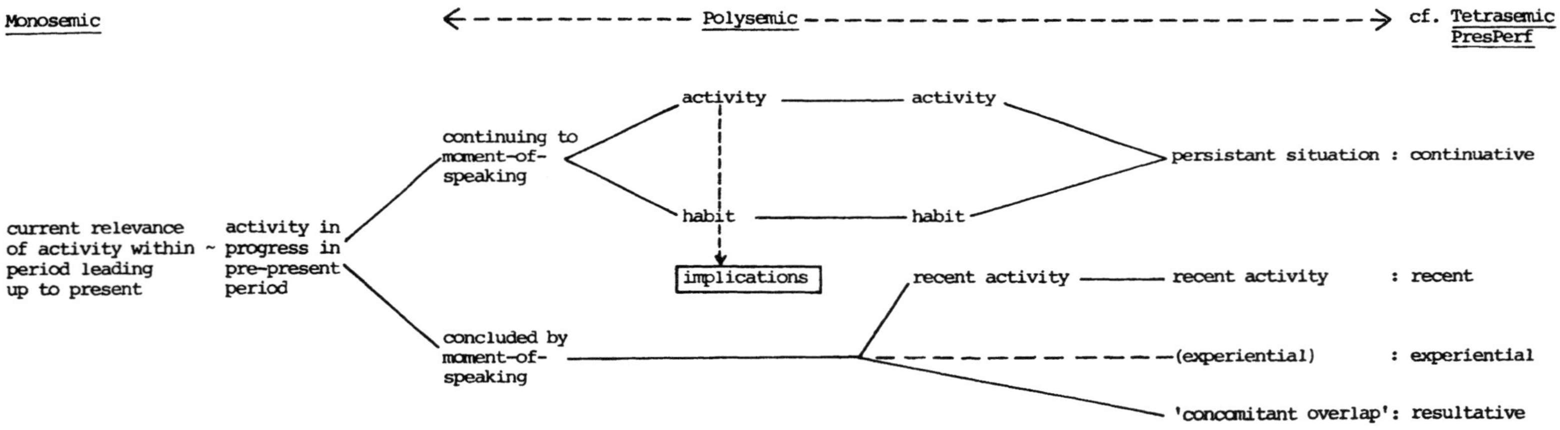

Table III

Standard English	Meaning/Reading	Hiberno-English
Preterite	Then time	Preterite
Perfect	Indefinite Anterior	Preterite
Perfect	Hot News	BE after V-ing
Perfect	Resultative	HAVE NP V-en (dynamic) BE V-en (mutative)
Perfect	Extended Now time	Present
Present	Now time	Present

Definitions

Indefinite Anterior:	Event occurring at unspecified time in period leading up to present.
Hot News:	Event located at point separate from but close to present.
Resultative:	Past event with present relevance.
Extended Now:	Situation initiated in past that persists into the present.

Examples

Indefinite Anterior:	StEng.	I've only seen ET once.
	HibEng.	I never saw a gun in my life nor never saw a gun fired.
Hot News:	StEng.	Al's just arrived.
	HibEng.	A young man's only after getting shot out there.
Resultative:	StEng.	Peggy has swallowed an orange.
	HibEng.	She's nearly her course finished.
Extended Now:	StEng.	I've known Sam for some time.
	HibEng.	I know his family all my life.

eNE	Irish	Meaning	Hiberno-English
Preterite	: Preterite	Then time Indefinite Anterior	Preterite
--	Cop S tréis (O) V -	Hot News	BE after V-ing
BE V-en HAVE NP V-en	Cop S PastPrt Loc -	Resultative	BE V-en HAVE NP V-en
Present	: Present	Extended Now time Now time	Present

NB. The above is taken, with minor modifications, from Harris 1984.

interpretation of 'resultative' (which is also not without its pragmatic problems, failing a 'perfecto-praesentium' interpretation) most events reported as 'hot news' will have effects on the present (i.e. the current) situation. And again, the 'experiential' perfect is, strictly speaking, the ascription of a property to an individual or set of individuals rather than the meaning of the Perfect; and a 'resultative' perfect (which may be identical in form) is, strictly speaking, a description of the world subsequent to the occurrence of an event. Thus, *He's seen ET* (without adverbial modification (cf. 2.1.1) can, depending on context and to an extent on intonation, be understood as an ascription ('a member of the set of those who have seen ET') and a description ('accomplishment of the task of seeing ET').[5]

The second reason is that determining how linguistic system L_1 (e.g. Standard English) refers to an object-world situation in contrast to linguistic system L_2 (e.g. Hiberno-English) is establishing a pragmatic equivalence, not a semantic one. A semantic equivalence, which is probably very rare in practice, obtains when two forms 'say' the same thing, i.e. they have the same denotation (or intension), entailments, implications and presuppositions.

What Harris's correspondeces might show, under ideal conditions, is that

---

5 We may note in this connection that McCawley himself points out the pragmatic nature of differentiating 'existential' (alias 'experiential') from 'hot news' readings:

Denis Brain has died in an auto accident (hot news).
Denis Brain and many other famous musicians have died in auto accidents. (existential)

We may also point out that the claim has been made (Kilby 1984:27-28) that the 'indefinite past' or 'experiential' reading is often linked to a stressed perfect auxiliary. If this were generally so, it would be evidence for a quasi-grammatical distinction. But closer inspection reveals that stressing the auxiliary *have* is dependent on the usual focusing conditions. Thus:

I háve opened it (= the window)

may have an experiential reading, but the stressing of *have* is determined by the contrast with the current situation in which the window is seemingly unopenable. The following, too, should demonstrate that the reading is dependent on the context, and that stressing plays no part in it whatsoever.

-- Do they still use poisoned darts?
-- Oh, ask George about that. He's been to Malaya. (indefinite ~ experiential)

-- My God, George is brown!
-- Isn't he? He's been to Malaya. (recent indefinite)

-- How's George's publicity tour going.
-- Fine. He's been to Malaya. (fulfilment ~ achievement)

the meaning of a form in L_2 (e.g. of the BE + *after* + V-*ing* construction) is included in the meaning of a form in L_1 (e.g. Standard English Perfect). On the other hand, the fact that both Standard English and Hiberno-English have forms like: *She has the boat sold* with a measure of synonymy but widely differing frequency of application to similar situations suggests that the applicability of 'meaning inclusion' may be spurious.

1.4 Support for the monosemic view

We may approach the question from another angle, that of linguistc methodology. Most of the polysemic (or implicitly polysemic) accounts fail to take proper account of minimal distinctivity; they posit uses of PresPerf which involve different contextual conditions rather than minimal distinctivity. In other words, they ignore the ceteris paribus principle in the establishment of linguistic categories. Thus, although most writers note that the continuative PresPerf obligatorily co-occurs with an adverbial of time duration, they still cast their statements about PresPerf in the following way: the (continuative) Present Perfect is used to denote an action, etc. that continues up to the moment of speaking. This kind of description is factually inadequate, descriptively misleading (e.g. for language teaching) and linguistically improper. The kind of formulation needed, allowing that other inadequacies may still be present, is something like: the Present Perfect is used *in combination with an adverbial denoting duration of time*, etc. to refer to an action, etc. that continues up to the moment of speaking. In other words, it is not the PresPerf but the adverbial that lies behind the continuativity. This point has been argued in greater detail and much more vociferously in Sørensen 1964.[6]

Though Sørensen himself upholds the monosemic analysis of PresPerf, this is not directly proved by his insistence that the adverbial is responsible for the various readings of PresPerf. And there is a little more to it than adverbials. As Schopf has repeatedly pointed out (Schopf 1976:3, 1984:13, and elsewhere) the reading of a predication is dependent on the interaction of tense, aspect, temporal adverbial and 'event-notion' (Geschehenskonzept), and a similar, though less precise point is made in Twaddell 1963: "compatible by-products of the semantic combination of context, situation, the lexical verb,

6 Not all Sørensen's claims can be accepted - a number of his examples strike the present author as at best idiosyncratic - but his demonstration that the continuative and resultative perfect can be accounted for fully by temporal and locative adverbials is thoroughly convincing.

and the 'Current relevance' signal" of the Perfect (ibid.:9). If the nature of the lexical verb, or more precisely, the 'event-notion' of a lexical verb is taken into account, we can arrive at certain minimal contrasts in the reading of PresPerf. Two fairly familiar cases are illustrated in:

I've got a book.	('perfecto-praesentium')
I've received a book.	('resultative')
John's gone to Wales.	('resultative')
John's been to Wales.	('experiential')

However, the readings suggested by the glosses are by no means incontrovertible. Even if they were incontrovertible, the monosemic position is only disproved if we can show that differences in reading cannot be explained without residue by reference to properties of the 'event-notions' of the lexical verbs (or better, the predicates) involved.

2. The 'isolated' Perfect

2.0

One way to examine the plausibility of the 'monosemic' position is to test just how ambiguous PresPerf and PresPerfProg forms are when they appear in effective isolation - 'effective' because it is rarely possible to completely exclude the context-of-situation, let alone the context-of-discourse. For practical purposes, we will consider an 'isolated' perfect to be one that is without temporal adverbial specification and deictic elements linked exclusively to S (point of speech) in immediate linguistic context.

It could be said that the 'hot news' PresPerf is a case par excellence of an 'isolated' perfect, since it is typically the introductory announcement of a report. However, 'hot news' has more to do with the semantics of discourse than the semantics of tense. If we consider the somewhat gruesome opener to a news broadcast, used recently as a clip in the satirical ciné-montage "Atomic Café":

Julius and Ethel Rosenberg have gone to the Electric Chair.

then we can see that it is a 'hot news' use in terms of text, but equally a resultative perfect in view of the fact that the cryptophoric expression: *go to the Electric Chair* means 'to be executed on the Electric Chair' and when cast in PresPerf entails 'be dead'. It should be pointed out that a 'hot news' reading has not been posited for PresPerfProg, but clearly:

Mrs Thatcher has been visiting the West Country

though presenting an activity rather than an event, fulfils the discourse conditions for 'hot news'. It could equally, given an extra-linguistic context of unusually large contingents of police, be a case of 'concomitant overlapping'. Linguistically, there does not seem to be any real ambiguity about what the utterance asserts.

2.1 The tests

In order to discover whether there is any preferred reading typically associated with 'isolated' perfects and whether there is a typical link between preferred reading and 'event notion', an examination was made of a relatively large number of predications. These were divided into two sets: Set Q (based on predicates listed in Quirk et al. (1985:201) and subdivided according to their 'situation types'), and Set Sch (based on predicates cited in Schopf (1984: 89-116 and 235-250) to exemplify the interaction of his event-notions and aspect) and subdivided according to Schopf's 'event-notions', which are only partly isocategorical with Quirk et al's 'situation types'. These sets, together with the test results, are presented in Table IV (p.130ff.). Examples cited in the following discussion have also been taken from pilot tests with other sets of predicates.

We may note, before discussing the tests, that Crystal 1966, to whom we shall have cause to refer in more detail later (3.1), establishes (on the basis of corpus tests) the following as meanings of PresPerf and PresPerfProg when not adverbially specified:

"action seen as recently completed, in unspecified past, removed from present

i. duration not emphasized (i.e. PresPerf) ...
ii. duration emphasized (i.e. PresPerfProg) ...

action seen as very recently completed, in unspecified past, results usually perceivable

iii. duration not emphasized (i.e. PresPerf) ...
iv. duration emphasized (i.e. PresPerfProg) ..."

(Crystal 1966:26)

<u>Table IV</u>

Set Q (based on predicates listed in Quirk et al. 1985)

			PresPerf	progr.^{able}	PresPerfProg
A.	1.	Jane BE tall	e	⋆	-
	2.	Man HAVE two legs	(e)	⋆	-
	3.	A whale BE a mammal	(e)	⋆	-
B.	4.	Tom BE angry	i	→	→ a
	5.	Fred THINK that pigs fly	e	→	t
	6.	Mary BE ill	i	⋆	-
	7.	Fred LOVE his mother	e	→	→ a (cf. f!)
	8.	Martita OWN a stallion	e	⋆	1
	9.	Pete RESEMBLE an ape	e	?	→ 1
C.	10.	Fred LIVE in Paris	e		t
	11.	Sasha STAND on one foot	f		a
	12.	Dick LIE on the floor	f		a
	13.	Jane SIT on the bed	f		a
D.	14.	It RAIN	o		a
	15.	It SNOW	o ~ r		a
	16.	The water BOIL	f ~ r		a
	17.	The sun SHINE	o		a
	18.	The fire GLOW	(f)		a
E.	19.	Mary DRINK	(f)		a
	20.	Papa SEW	f		a
	21.	Grandma WRITE	f		a
	22.	The neighbours HUNT	e		a
	23.	Mrs Shufflewick TALK	f		a
	24.	The girls PLAY patience	f ~ e		a
F.	25.	The tomatoes RIPEN	r		p → t
	26.	Jocelyne GROW UP	r		p
	27.	Sue's health IMPROVE	r		p
	28.	The components SEPARATE	r		p → s
	29.	The sky TURN RED	r		p → t
G.	30.	Grandma WRITE a novel	r ~ f		a
	31.	Ian EAT hamburger	f		a
	32.	Mary DRINK a cocktail	f		a
	33.	Clive FILL UP Ma's glass	r		a → t
	34.	Mme Curie DISCOVER radium	r ~ f		a p
H.	35.	Martin SNEEZE	o		s
	36.	The spaceship EXPLODE	r		a
	37.	Silas BLINK	o		s
	38.	A light FLASH	o		s
	39.	The ball BOUNCE	o		s
I.	40.	Victor TAP on the window	f		s
	41.	Celia NOD her head	f		s
	42.	The sentry FIRE a warning shot	f		- (cf. his gun)
	43.	Debbie KICK her brother	f		s
J.	44.	The price DROP	r		p
	45.	June RECEIVE an invitation	r		a
	46.	Quint CATCH the flu	r		a p
	47.	A Boeing TAKE OFF	r ~ f		a
	48.	The plane ARRIVE	r		a
	49.	The Duke DIE	r		p
K.	50.	Bill SIT DOWN	r		a
	51.	A spectator CATCH the ball	r		s
	52.	Vi SHOOT a tiger	r		a
	53.	David BEGIN a novel	r		a
	54.	Basil STOP the car	r		a → s

Set Sch (based on predicates in Schopf 1984:89-116, 235-250)

			PresPerf	progr.able	PresPerfProg
A	1.	BE cultured	e	→	→ a
	2.	BE musical	e	→	→ a
	3.	BE tall	e	*	-
	4.	BE short	(e)	*	-
	5.	BE intelligent	e → i ~ f	→	→ a
	6.	BE a giant	e	→	→ a
	7.	BE German	e	→	→ a
B	8.	BE asleep	i ~ c	*	-
	9.	BE tired	i	→	(→ a)
	10.	BE dizzy	i	*	-
	11.	BE hungry	i	*	-
	12.	BE dirty	i → f	→	→ a
	13.	BE happy	i	→	(→ a)
C	14.	BE friendly	i	→	→ a
	15.	BE naughty	i	→	→ a
	16.	BE kind	i	→	→ a
	17.	BE brave	i ~ e	→	→ a
D	18.	BE red in the face	i	*	-
E	19.	CONSIST OF alcohol	(e)	*	→ 1
	20.	OWN a car	e	?	→ 1
	21.	POSSESS a firearm	e	*	-
	22.	CONTAIN drugs	e ~ i	*	→ 1
F	23.	KNOW her/the truth	e/i	*	-
	24.	BELIEVE her/that + Clause	i → f	→	(→ t) → p
	25.	THINK that + Clause	e	→	→ a
G	26.	LOVE her/the play	e/i	→	→ a/→ t
	27.	HATE her/the play	e/i	→	→ a/→ t
	28.	DETEST her/the play	e/i	→	→ a/→ t
H	29.	SEE her	e ~ i → f	→	→ a → t
	30.	HEAR her	e ~ i(→ f)	→	→ t
I	31.	(My feet) HURT	(i)		t
	32.	(My head) ACHE	(i)		t
	33.	(My nose) ITCH	(i)		t
	34.	(The blanket) TICKLE	(i)		t
	35.	(My eyes) SMART	(i)		t
J	36.	LIE on the floor	i ~ f		t ~ a
	37.	STAND in the corner	i ~ f		t ~ a
	38.	LEAN against it	i ~ f		t ~ a
K	39.	RULE England	e (→ f)		1
	40.	RAISE cattle	e ~ i ~ f		a ~ t
	41.	GROW potatoes	e ~ f		a ~ t
L	42.	SMOKE	e ~ f		t ~ a
	43.	GO to school	r		t
	44.	PLAY chess	e ~ i ~ f		a
	45.	DRIVE a cab	e ~ f		t ~ a
M	46.	SMOKE	f		a
	47.	DANCE	f		a
	48.	BREATHE	f		a
	49.	PUSH a cart	f		a

			PresPerf	PresPerfProg
N	50.	SNOW	r	a
	51.	HAIL	i	a
	52.	RAIN	o	a
	53.	DRIZZLE	i	a
	54.	FREEZE	r	t
	55.	(The storm) ROAR	(i)	a
	56.	(The sea) RAGE	(i)	a
	57.	(The sun) SHINE	i ~ o	a
	58.	(Time) PASS	i ~ o ~ f	t
O	59.	RUN	f	a
	60.	WALK	f	a
	61.	SING	f	a
P	62.	LAUGH	f	a
	63.	BLUSH	r	s
	64.	GROAN	f	s ~ a
	65.	ROAR	f	s ~ a
	66.	GRIN	f	a
	67.	HOWL	f	s ~ a
	68.	SMILE	f	a
Q	69.	KNOCK	f	s
	70.	TAP	f	s
	71.	SNAP at him	i ~ f	s
	72.	NOD one's head	f	s
	73.	HICCUP	i ~ f	s
	74.	SNEEZE	i ~ f	s
	75.	COUGH	i ~ f	s
	76.	BELCH	i ~ f	s
	77.	SLAP her	f	s
	78.	CLAP someone on the back	f	(s)
	79.	BLINK	i ~ f	s
	80.	TICK	i ~ f	s
	81.	CROSS oneself	f	s
Q'	82.	(The door) BANG	i ~ f	s
	83.	(The latch) CLICK	i ~ f	s
	84.	(The rifle) CRACK	i ~ f	s
R	85.	SWITCH ON/OFF the light	r	a ~ t
	86.	SNAP OPEN/SHUT	r	s ~ t
	87.	OPEN the window	r	a ~ s
	88.	BANG SHUT	r	s
	89.	BLOW OPEN	r	s
	90.	CRACK	r	p
	91.	SPLIT	r	p
	92.	BREAK	r	p
R'	93.	LOOK AWAY	r	a
	94.	LOOK UP/DOWN	r	a
	95.	LOOK OUT of the window	f	a
	96.	LIE/SIT DOWN	r	a
	97.	SIT UP	r	a
	98.	LEAN BACK	r	a
	99.	LEAN OUT of the window	r	a
	100.	BEND FORWARD	r	a
R''	101.	BEGIN to rain	r	?
	102.	STOP talking	r	?
	103.	START to snow	r	?
	104.	(The noise) STOP	r	p

			PresPerf	PresPerfProg
S	105.	GROW OLDER	r	p
	106.	(The universe) EXPAND	r	p
	107.	(Traffic) INCREASE	r	p
T	108.	IMPROVE	r	p
	109.	INCREASE	r	p
	110.	DIMINISH	r	p
	111.	DETERIORATE	r	p
	112.	GROW BIGGER	r	p
	113.	BECOME OLDER	r	p
	114.	BECOME SMALLER	r	p
	115.	ACCELERATE	r	a ~ p
U	116.	(The hay) DRY	r	a ~ t
V	117.	DRAW a circle	r	a
	118.	WRITE a letter	r	a
	119.	RUN a mile	f ~ r	a
	120.	EAT an apple	f ~ r	a
	121.	KNIT a sweater	r	a
	122.	BUILD a house	r	a
	123.	SMOKE a cigarette	f ~ r	a
V'	124.	WASH the car	r	a
	125.	PAINT the wall	r	a
	126.	CLEAN the room	r	a
W	127.	WIN a race	r	a
	128.	DEFEAT the enemy	r	a ~ p
	129.	FIND a room	r	a
	130.	REACH the station	r	(t)
	131.	ARRIVE at the station	r	t(~ a)
	132.	DISCOVER the virus	r	a ~ p
	133.	REACH the hilltop	r	(t)
	134.	(The plane) TAKE OFF	r	a

Key: e = experiential ~ remote indefinite
i = recent indefinite
o = occurrence
f = fulfilment ~ achievement
r = resultant state
() = if interpretation at all possible
/ = alternative interpretation

a = activity
p = incremental process
t = temporary habit ~ recent limited state-of-affairs
s = series of occurrences
l = limited state ~ recent limited state

⟶ = change of sense
⋆ = not progressivizable

Key to Groupings

	States	PresPerf	PresPerfProg
A	durative	exp.	act.
	(incl. co-extensive)	(if Prog, sense change)	
B	temporary	indef.	act.
		(if Prog, sense change)	
C/D	quality/behaviour	indef.	act.
		(if Prog, sense change)	
E	relation	exp.	lim.stat.
F	mental	exp.	temp.hab./act.
		(if Prog, sense change)	
G	emotional	exp.	act./lim.stat.
H	perceptual	ful.	lim.stat.
I	bodily sensation	(indef.)	act.
J	locative manner	ful.	act.
K	potential ('generic states')	exp.	lim.stat.
L	habitual ('specific states')	exp.	lim.stat.
	Simple Process		
M	activity	ful.	act.
N	process	(!) res./indef./ful.	act.
O	Initiated process	ful.	act.
P	Quantified process	ful.	act.
Q,Q'	Punctual event	ful.	ser.
R	Punctual change	res.	ser./act./incr.
	R' into process	res.	act.
	R'' aspectual	res.	?/incr.proc.
S	Non-quantified incr.proc.	res.	incr.proc.
T	Initiated incr.proc.	res.	incr.proc.
U	Quantified incr.proc.	res.	act.
	Accomplishment		
	V effective	res.	act.
	V' affective	res.	act.
W	Achievement	res.	act./incr.proc.

The results of Crystal's tests can be very broadly upheld,[7] but, as will transpire, the differentiation of the 'recent' from the 'very recent' is pragmatic (and actually results from Crystal's overstatement of the interaction of tense and adverbials).

2.1.1 Tests with PresPerf

On the basis of an initial test with Set Q predications, the results of which were relatively uniform, the following hypothesis was made:

1. If a predication is non-dynamic, then the 'isolated' PresPerf will denote the terminated state and locate it prior to the moment of speaking (S) (indefinite reading).

    a. if the state is temporary, the temporal location is proximal (recent indefinite).
    b. if the state is non-temporary, the temporal location is distal (experiential/remote indefinite).

(This reflects an inherent duration of states rather than Quirk et al's differentiation between qualities and states. Generic propositions are notoriously resistent to even an 'experiential' reading of PresPerf.)

2. If a predication is dynamic (including Quirk et al's 'stance' type), then the 'isolated' PresPerf will denote the complete event and locate it prior to S.

    a. if the event is non-conclusive and
        i. non-agentive, the implicature is completion.
        ii. agentive, the implicature is achievement/fulfilment.
    b. if the event is conclusive, the implicature is resultant state.

(This does not preclude an 'experiential' reading, since, as suggested above, ascription is a part of discourse semantics, not tense semantics).

This hypothesis was tested with Set Sch. It was assumed that Quirk et al's 'conclusive' would correspond to Schopf's event-notions including t_f and resultant state (quantified processes, punctual changes, quantified incremental processes, accomplishments and achievements) and non-conclusive with those excluding t_f (all others). (Schopf's event notions do not take explicit account of agentivity and causality.) The tests were conducted on

---

7 There are, for instance, one or two highly questionable collocations predicted by Crystal's analysis, e.g. PresPerf with Crystal's C3a adverbials, viz. S(pecific) T(ime) *ago/back*, etc. would predict the acceptability of **I've seen her some time ago* and **I've seen her a week yesterday*.

the 'default' principle, i.e. unless the expected values were clearly at odds with one's intuitions, they were considered to hold, even if the assignment was not always unequivocal.

The unexpected results have been indicated in the tables. Some of them deserve comment.

Although it might be expected that initiated incremental processes like *improve* and *expand*, because they do not entail a t_f, should not have a resultant reading, they quite clearly do. The resultant state derives from the point at which the incremental process was halted. It seems therefore that 'directionality' or 'incrementality' should be taken together with having t_f for a resultant state. On the other hand, quantified (non-directed) processes like *smile* seem to favour a fulfilment or achievement reading, thus having a t_f per se does not seem to entail a 'resultative' perfect reading. An exception to this is *blush*, but it may well be that, at least as far as PresPerf is concerned, *blush* is a punctual change with resultant state rather like *look away*.

Predications involving bodily sensation resist PresPerf (but not PresPerf-Prog) unless specified by a (durative) adverbial (in which case the reading is continuative). At best they could have a 'recent indefinite' reading.

Predications involving (non-telic) activities (*paint*) or co-extensive behaviour (e.g. *work in a factory*) seem to favour either 'experential' or 'fulfilment' readings. The choice seems to be pragmatically determined. Thus *work in a factory, smoke, drive a cab* will normally sui generis prefer an 'experential' reading, while *play chess, dance, sing* will sui generis prefer a 'fulfilment' reading.

Continuous processes like *represent someone's interests, orbit the earth, beat out the rhythm*, though state-like in event notion, seem to favour a 'fulfilment' reading rather than a 'recent indefinite' reading. This must be attributable to their inherent dynamism.

Simple processes such as *snow, hail, rain*, etc. behave rather erratically - and again the factors seem to be pragmatic rather than to lie in their inherent temporal-structure (event notion). *Snow* and *freeze* produce obvious and lasting physical effects, *rain* to an extent, but *hail* and *drizzle* focus on the variety of precipitation, with *rain* presumably as hyperonym, resulting in no particular distinctive physical state. When conceived more in terms of process completion, the reading seems to be more 'fulfilment'; when conceived in terms of conclusion of activity, 'recent indefinite'. Arguably, these are pragmatically determined variants that can be subsumed under 'occurrence of the complete process in anterior time zone'.

We will not pursue individual results of this test further. Many of the exceptions to predicted behaviour can be explained pragmatically; one or two indicate that the dynamic vs. non-dynamic distinction is relevant (since ambivalent predictions pattern ambivalently); and a few, indeed, may be attributable to erroneous classification.

Table V (p.138f.) shows the correlation of PresPerf readings with (a) Quirk et al's 'situation types' and (b) Schopf's 'event notions'. It can be seen that the better correlation is with Quirk et al's system, i.e. that agency and 'conclusiveness' seem to be reasonable indicators for predicting PresPerf readings. Ironically enough, Quirk et al's system was designed to explain progressivizability!

2.1.2 Tests with PresPerfProg

The set of impressionistic names for readings of PresPerfProg (cf. Table II) is somewhat obscurer that that for PresPerf. The term 'activity' seems to encompass a great number of interpretations, but simply because it denotes a dynamic state-of-affairs without any indication of its composition or conclusivity. We will attempt to classify readings of PresPerfProg initially in terms of: 'activity', 'serial activity', 'incremental process', 'temporary habit' and 'limited state'.

An 'activity' is, as suggested above, a dynamic state-of-affairs irrespective of whether the dynamism is associated with agency or not. We will restrict it to a state-of-affairs that is continuous, not intermittant. A 'serial activity' is a non-continuous activity involving a succession of actions or sub-events. An 'incremental process' is a non-agentive, constantly changing directed activity. A 'temporary habit' is a temporally limited state-of-affairs involving intermittent actualization. And a 'limited state' is a (recent) inherently temporally limited non-terminated state. It is arguable that the distinction between the last two is unnecessary.

An initial test of readings of PresPerfProg in Set Q in terms of the above revealed the following pattern.

1. Qualities could not appear in PresPerfProg.
2. Non-temporary states (i.e. those with an experiential reading in PresPerf) may have
   i. a 'temporary habit' reading if cognitive.
   ii. a 'limited state' reading if relational.

Table V(a)

Quirk et al's situation types and PresPerf.

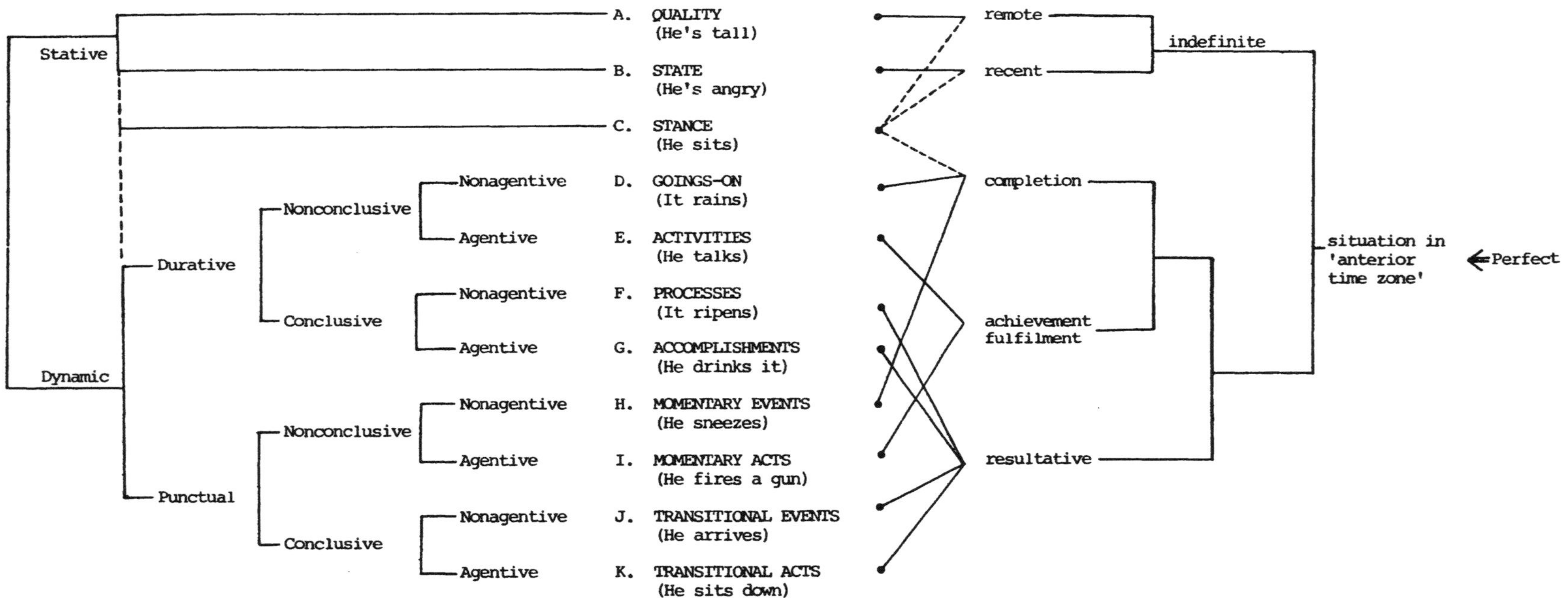

<u>Table V(b)</u>

<u>Schopf's event notions and PresPerf</u>

STATES	durative/coextensive	(BE tall)
	temporary	(BE asleep)
	quality/behaviour	(BE friendly)
	relation	(CONTAIN drugs)
	mental	(KNOW the truth)
	emotional	(LOVE her)
	perceptual	(HEAR her)
	bodily sensation	(HURT (of <u>feet</u>))
	locative manner	(LIE on the floor)
	habitual	(DRIVE a cab)
	potential	(RULE England)
SIMPLE PROCESSES	actional	(DANCE)
	processual	(RAIN)
INITIATED PROCESSES		(WALK)
QUANTIFIED PROCESSES		(SMILE)
PUNCTUAL EVENTS	actional	(KNOCK on the door)
	processual	(BANG (of <u>door</u>))
PUNCTUAL CHANGES	of state	(SWITCH OFF the light)
	transition into process	(LOOK AWAY)
	aspectual	(BEGIN to rain)
INCREMENTAL PROCESS		(GROW OLDER)
INITIATED INCREMENTAL PROCESS		(INCREASE)
QUANTIFIED INCREMENTAL PROCESS		(DRY (of <u>hay</u>))
ACCOMPLISHMENT	effective	(DRAW a circle)
	affective	(WASH the car)
ACHIEVEMENT		(WIN a race)

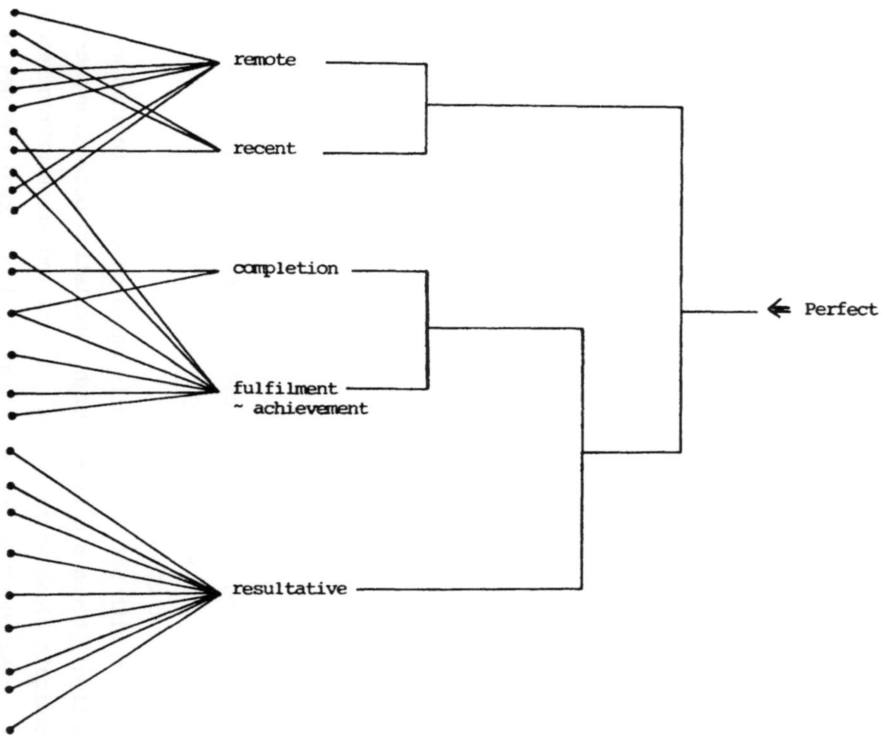

iii. an 'activity' reading with a switch from non-dynamic to dynamic.
3. If the predication is non-conclusive, PresPerfProg will have
   a. an 'activity' reading, if non-punctual.
   b. a 'serial activity' (iterative) reading, if punctual.
4. If the predication is durative and conclusive, PresPerfProg will have
   a. an 'activity' reading, if agentive.
   b. an 'incremental process' reading, if non-agentive.
5. If the predication is punctual conclusive, PresPerfProg will have either an 'activity' or an 'incremental process' or a 'serial activity' reading dependent upon non-obvious factors.

(It should be added that 'stance' predications that would not normally have an 'experential' reading in PresPerf behave like Quirk et al's non-conclusives, while <Fred LIVE in Paris>, which would have an 'experiential' reading in PresPerf, behaves like a non-temporary state.)

In addition to tests parallel to those for PresPerf, a check was also made on whether predications were 'progressivizable', i.e. could appear in Present Progressive. Those predications that resist 'progressivization' or allow it only with a sense change coincide more or less exactly with those that do not occur in PresPerfProg or do so only with a sense change. An apparent exception is <OWN a car>, which is not clearly excluded as a PresPerfProg, but is as a PresProg, cf. 3.4.0. This might lend credence to the 'aspectual modification view' (but cf. below).

The pattern with Set Sch, which is based on a more refined analysis of 'event notion', reveals a fairly high correlation between event-notion (or at least predicate class within event notions) and PresPerfProg reading, except in three cases: punctual changes, 'aspectual' punctual changes and 'achievements'. Much less satisfactory is the correlation between PresPerf reading and PresPerfProg reading. The pattern is presented following Set Sch in Table IV.

A cross-check with Schopf (1984:235-50) reveals that the terms in which the classification of readings has been made reflect a subset of the notions attached to the non-perfect progressive. An 'activity' reading of PresPerf-Prog corresponds to a focusing on the process phase or pre-phase or post-phase by a non-perfect progressive. An 'incremental process' corresponds not just to a process phase but to what is essentially a non-agentive mechanistic directed process phase. And finally, a 'temporary habit' and a 'limited state' correspond to a state-of-affairs that is temporally bounded.

2.1.3 Discussion

We may attempt an explanation of at least some of the poor correlations.

While *switch off/on the light* and *open the window* do not differ very much from *snap/open/shut* or *bang shut* in terms of inherent temporal structure, they do differ pragmatically in terms of how intrinsic the action is to the referent of the subject: *switch off/on the light* and *open the window* are more extrinsic in that they pragmatically entail taking up a position in order to perform the action, which means that there is a phase prior to the named event that can be selected through the choice of progressive aspect. And while *bang shut* and *blow open* are effectively instantaneous, *crack, split,* and *break* can be instantaneous or incremental.

With achievements, or what have been classed as achievements, pragmatic differences can also be found. *Win a race* and *find a room* involve a presupposed process phase (running in a race, looking for a room) which is in a privative opposition to the post-phase (having won a race, found a room), while *defeat the enemy* and *discover the virus* involve gradable process phases (the enemy can be defeated in stages, the virus can be discovered by a process of gradual elimination - one suspects its existence before isolating it). Cast as PresPerfProg, *win a race* and *find a room* strongly suggest achievement (despite the progressive form), while *defeat the enemy* and *discover the virus* do not. Locative achievements like *reach/arrive at the station*, while seeming to presuppose a goal-directed process, do not select this phase or the general activity when cast in PresPerfProg, unlike *arrive* without locative expression, which in PresPerfProg denotes the general activity. The reasons for this are not clear.

We may conclude this discussion by noting that while there is a certain correlation between the readings of PresPerf and those of PresPerfProg with states, the more important factor is whether progressive aspect is possible and if it is, with what sense change. And as far as the non-states are concerned, it seems clear that the PresPerfProg reading cannot be directly predicted from the PresPerf reading. A better, though not infallible, prediction can be made if we know the nature of the phase of the event notion that is selected by progressive aspect in PresProg. In other words, it is more useful to see PresPerfProg as a perfecticization[8] of PresProg, than to

---

8 This term is motivated by Bauer's 1970 choice of 'perfectic' to avoid unwanted confusion with 'perfective', which is, rightly, reserved for an aspectual distinction.

see it as a progressivization of PresPerf.

2.2 The modified Reichenbachian approach

In the light of the foregoing, we should now consider how PresPerf and PresPerfProg may be represented. The proposal made in Reichenbach (1947:290) for PresPerf in terms of three time points: S (point of speech), E (point of event) and R (point of reference) in a configuration:

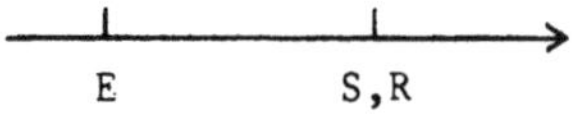

has been remarkably seminal: despite numerous criticisms and modifications, the essence of Reichenbach's proposal, that the point of reference is simultaneous to the point of speech with the point of event anterior, in contradistinction to the Past, where point of reference and point of event are simultaneous and both anterior to the point of speech:

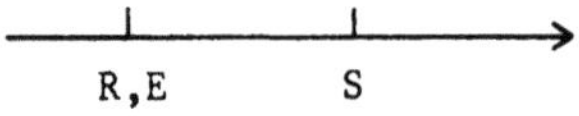

has been frequently reaffirmed. This is attributable largely to the fact that this configuration gives a formal specification of the essential presentness of PresPerf:[9] it fits into modes of description and narration that are present-oriented, rather than modes that are past, i.e. oriented in identified or identifiable anterior time.

It will not be possible to examine in detail the ways in which Reichenbach's formal proposal has been criticized, found wanting and modified.[10] Instead, we shall take the configurational types used in Schopf (1984:320-328) as a basis. These, as Schopf notes, involve a temporal frame extending from the past up to the moment of speaking, which includes the event, thus permitting differentiation from configurations for PresPerfProg. Schopf does not, however, commit himself to the Bennett/Partee 1978 proposal that "the present perfect always involves a reference to a definite interval of time which starts at some point in the past and includes the present moment, or the

9 This is the basis upon which the term 'Present Perfect' has replaced the older 'Perfect' (and analogously 'Past Perfect' has replaced 'Pluperfect') in most grammars.

10 The subject of reference time or point of reference is the topic of another essay in this volume, viz. Cornelia Hamann: 'The awesome seeds of reference time', p. 27-69. Cf. also discussion of Reichenbach and logical approaches in Schopf (1984:203-208 and 212-223).

moment of utterance." (Bennett/Partee (1978:8), quoted in Schopf (1984:322)). The wording of Bennett/Partee's proposal is problematic in that the 'reference' is not to a definite interval (except when the PresPerf collates with an adverbial referring to a definite interval), though it might be reference to a point terminating an interval (cf. below).

Given this kind of iconic representation, the "essential unity" (Sørensen 1964) of the PresPerf can be brought out on the one hand, while keeping track of the essential diversity of event notions, on the other. In Table VI (cf. p.144) a selection of event notions has been represented in Pres and PresPerf and PresProg and PresPerfProg. What perfecticization does is push the event (as a whole, in the case of PresPerf) or a phase thereof (in the case of PresPerf-Prog) into what Quirk et al 1985 refer to as the "anterior time zone", i.e. a zone which extends no further forward in time than the point of reference (R) which is simultaneous with the point of speech (S).

It might be said that the R of Pres and PresProg "homes in" on whatever state-of-affairs obtains in the object world, or sets a reference for an event in the object world, whereas the R of PresPerf and PresPerfProg "takes stock" of the state of the object world, and thus sets a temporal limit (S) on the object world, not directly on the state-of-affairs or event. It is in this sense, perhaps, that Bennett/Partee's "reference to a definite interval" should be interpreted. If there is a starting point, then (without adverbial specification) it can only be an implicature from the existence in time of the participants in the object-world situation: it is not referred to.

In claiming that 'perfecticization' shifts the event or the phase of the event or the state-of-affairs into an "anterior time zone", two caveats are necessary. First, with non-coextensive states in PresPerf, it is only the state that is asserted as being anterior to R. Thus, *she's been asleep* is neutral as to whether the referent of *she* is awake at S or not. For pragmatic reasons *I've been asleep* will normally be interpreted as if the conclusion of the state is also anterior to R (= S). With co-extensive states, the conclusion also seems to be asserted. Second, in all cases where progressive aspect can select a pre-phase or the directed process phase, the activity reading of PresPerfProg will not exclude fulfilment of the event. Thus, *she's been winning the race* in answer to: *What's she been up to?* would certainly imply the achievement of <WIN the race>. The status of this is probably that of a conventional implicature, PresPerfProg only asserting the phases selected by progressive aspect. Thus it is not non-sensical to

Table VI

	S R	Perfect S R	
She's asleep	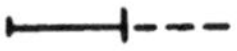	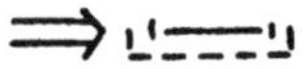	indefinite
She owns a car			experiential
She smiles		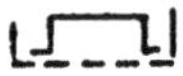	fulfilment
She sneezes		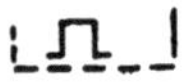	fulfilment
She switches off the light		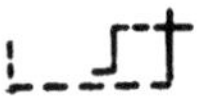	resultative
She grows older	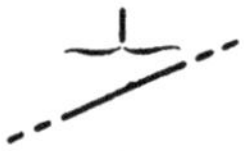		resultative
She knits a sweater		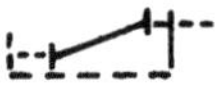	resultative
She wins the race	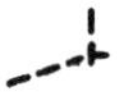		resultative
She's living in Paris	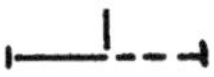		temp. habit
? She's owning a car			lim. state
She's smiling	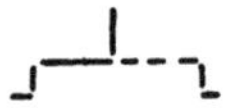	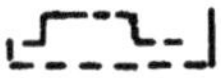	activity
She's sneezing	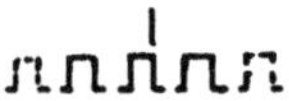		serial activity
She's switching off the light	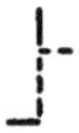		incept. activity
	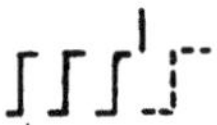	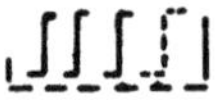	ser. activity/ temp. habit
She's growing older	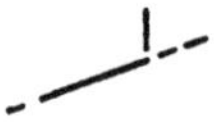	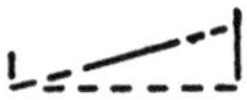	increment. process
She's knitting a sweater	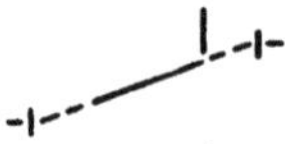	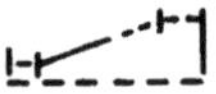 *	activity
She's winning the race	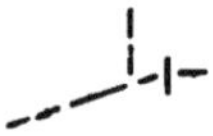	 *	activity

* the t_f phase, being outside the assertion, could lie to the right of R.

ask: *And has she done so?* in response to: *She's been switching off the light,* whereas it would be non-sensical in response to: *She's switched off the light.*

3.0 The collocated Perfect

In this section we will discuss the effects on the interpretation of PresPerf and PresPerfProg of collocation with temporal adverbials. But first we should take account of Crystal's 1966 findings on the meanings of tense and adverbial collocations.

3.1 Crystal's findings

3.1.1 Crystal establishes a list of temporal adverbials that can be used in potential response to *How often? How long?* and *When?* These are classified into 6 classes (A - F) and the classes into subclasses: A1 - 9, B1 - 5, C1 - 5, D1 - 4, F1 - 5. Certain common expressions such as *just* and *now* are omitted from the lists because of their anomalous patterning. Subclasses (including *now* and *just*) that are incapable of collocation with perfect tenses, or at least are non-attested in such collocations, are apparently few, involving only: *for the moment, for now, for the present/time being* (Crystal's B3); *for the future, from now on, henceforward, henceforth, hereafter, in (the) future* (B4); and *any moment ... now, a week/fortnight ... tomorrow, a week ... next Monday, in days ... to come, later this week ..., next time, (any/some day ...) next week ..., one of these days ..., some day, the day after tomorrow, the week ... after next, tomorrow (morning .../week/fortnight),* DAY OF WEEK/MONTH OF YEAR *next* (C1). Although these all involve futurity in some way, those of B3 can clearly collocate with perfect tenses, e.g. *She's switched it off for now/ the moment/the present/the time being.* On the other hand, there are numerous items that, according to Crystal's lists, should be collocatable, which in fact give unacceptable results: **She's switched it off once upon a time.* This may be a consequence of Crystal's approach to collocation.[11] He distinguishes three types of adverbial specification: i. adverbial in the same clause, ii. adverbial in a subordinate clause, iii. adverbial in remote context,

11 An alternative explanation is that Crystal's analysis of his data (or at least his presentation of it) is at fault.

governing the temporal specification of later clauses until further adverbial specification is used. It is highly questionable as to whether Crystal's types ii and iii should be treated as adverbial specification, though that is not to say that they may not set up a relational network which interacts with tense-choice and other temporal expressions. The difficulty involved in Crystal's approach can be illustrated by the following:

* I have seen him yesterday.

-- Have you seen him?
-- Yes, yesterday.

Within a clause, collocation of PresPerf with *yesterday* is unacceptable: beyond the clause, a specification of the time at which an indefinite event occurred by means of *yesterday* is perfectly in order. The respondent is in fact treating the question (which sets up a 'perfect' frame) as 'have you seen him, and if so, when?'. The *yes* affirms the occurrence of the event within a temporal frame presupposed (and shared) by speaker and hearer: the *yesterday* makes a new temporal reference, and is elliptical for: *I saw him yesterday*.

Though the discourse factor should not be underestimated, the temporal specification of Crystal's type i is of a very different order. And it is this kind to which this paper will restrict itself. Without a thorough understanding of how Crystal's type i specification works, examination of types ii and iii is idle speculation.

3.1.2 The meanings of the collocations

Crystal's analysis of the readings of PresPerf and PresPerfProg (which he does not explicitly differentiate in his classification) is, giving one example only for each of his classes, as follows:

Optional (or no) specification (cf. 2.1 above)

Action seen as recently completed, in unspecified past, removed from present

i. duration not emphasized: (*never, once, again, seldom, occasionally, often, usually, regularly*)
e.g. I've written three books; He's visited Rome; X has been ill; I've run 100 yards in 10 seconds; I've forgotten, etc.

ii. duration emphasized: (*for ages, some time ago* (!), *this morning*)
e.g. They've been building everywhere; It's been raining; Someone's been moving my papers; Who's been lighting matches, etc.

Action seen as very recently completed, in unspecified past, results usually perceivable.

iii. duration not emphasized: (*just*)
e.g. I've had a bath; You've been to town; He's given me this; I've cut my finger, etc.

iv. duration emphasized: (*just*)
e.g. We've been having a lesson; I've been finishing off our letters, etc.

Obligatory specification

v. past action, seen as recent, in specified time: *at the moment, lately, this morning, at dawn* (?), *already, all of a sudden, presently* (?), *at last, afterwards*.
e.g. Lately, I've been living in France.

vi. past duration, seen as recent (including periods of time inclusive of present): *all day, so far, for ages*.
e.g. I've lived here since ...; You've been reading about X for an hour? etc.

vii. past frequency of occurrence ('habitual'): *never, once, again, seldom, occasionally, often, usually, regularly; at the moment* (?), *lately, this morning, at dawn; already, almost (10 o'clock)* (!), *by now, a day earlier (than planned); at that age; all of a sudden, presently, at last, afterwards, after a few days*.
e.g. In due course, he's walked into the house (i.e. as we've expected).

viii. simultaneous perceivable activity: *almost*
e.g. He's almost reached the goal (in commentaries, etc.)

Perusal of the above will reveal a number of inconsistencies - some of the more glaring have been indicated by (?) or (!). A basic problem here seems to be the heterogeneity of some of Crystal's adverbial classes. In general terms, Crystal's study suffers, as many corpus-based studies do, from the inadequacy of the descriptive framework when applied to text, and even of the analyst's sensitivity to textual parameters.

3.2 The tests

In order to test the range of readings of PresPerf and PresPerfProg adverbial collocations, a much more modest set than Crystal's has been selected. It consists of:

again, all day, all his life, already, always, before, every day, for hours, for the moment, for years, hitherto, just, lately, meanwhile, never/not ... ever, now, occasionally, often, once, recently, regularly, seldom, since he was a boy, since 10 o'clock, so far, this morning, this year, until further notice, up to now, usually, (not) yet.

These have the advantage of being, on the one hand, relatively frequent in occurrence, and on the other, representative of (if not exhaustive of) Crystal's classes.

3.2.1 The function of the collocating adverbials

Before proceeding to the tests, we should examine what the communicative function (as opposed to the denotation(s)) of such adverbials is. It is a well-known fact that adverbials can occupy three basic positions in a clause: initial (I), medial (M) and final (F),[12] and that there is a tendency for certain adverbials to assume particular positions in the clause. In some cases, moreover, a particular reading of an ambiguous adverbial may be associated with a particular position (and in fact also with a particular configuration in prosodic patterning).

If we take as an example the syntactically ambiguous form *now* in collocation with the PresPerf of <he EAT it>, we can arrive at at least 8 readings:

1. Now, he's eaten it (, but that's no cause for alarm).
2. Now he's eaten it (, I'm not so worried).
3. Now he's eaten it (, whatever next?).
4. NOW he's eaten it (, 10 minutes ago he hadn't).
5. He's now eaten it (, as requested).
6. He's eaten it now (- there's not much we can do).
7. He's eaten it ... NOW!
8. He's eaten it NOW (- he hadn't earlier).

Readings 1 and 2 (a 'transitional discoursal conjunct' (cf. Quirk et al 1985: 636) and a subordinating conjunction (= *now that*) respectively) are not temporal adverbials: 3 to 8 are. The contrast in initial position is between a 'setting' (3) and a 'contrastive thematic' (4) use of *now*. In medial position (5) *now* is glossable as 'in the meantime' and presupposes the expected/required fulfilment of <he EAT it>. In final position, very much dependent on intonation, we have a 'summative' use, a 'simultaneous reference' use and 'contrastive rhematic' use. The positional and prosodic determination of readings of *now* aside, we can distinguish three uses of adverbial *now*: deictic (7), updating (3,6) and temporal contrasting (4,8).

In principle, we can envisage at least three relationships between an adverbial and a tense form: a where the tense form is dependent on the adverbial chosen; b where the tense form and adverbial are interdependent; c where the adverbial is dependent on the choice of tense. These three relationships may be termed: a 'adverb-dependency'; b 'mutual dependency'; c 'tense-dependency'.

This is not quite the same thing as saying that an adverbial is linked to a particular tense-form only or free to collocate with a variety of tense-forms.

---

12 For definition, illustration and refinement of these positions, cf. Quirk et al.(1985:490-501).

To take a fairly clear case of tense-dependency, *just* as an adverbial at M can collocate with Prog, Perf and PerfProg with effectively the same function (emphasizing proximity to R), but it does not specify R, nor the frame for the event, nor does it change the overall tense-configuration. Thus, there is a good case for claiming that *just* simply selects or focuses on a part of the denotation of the tense form: the reading of *He's just eaten it* is included in *He's eaten it*. The fact that *just*, unlike most other adverbials tested, cannot appear at I (or F with the same meaning, cf. *He's eaten it - JUST*, which comments on the degree of fulfilment (cf. *almost*) not the recency of fulfilment) supports this view. On the other hand, *since 10 o'clock*, which can hardly collocate with anything other than perfect tense forms (present or past) specifies the temporal frame for an event or state or activity. Here it makes sense to say that PresPerf or PresPerfProg is adverb-dependent. *Recently*, however, can be said to be in a mutual dependence relation with the tense form with which it occurs: with perfect tenses, the collocation as a whole denotes an event or state or activity within a recent period up to R (= S); with Past, the collocation as a whole denotes an event or temporary state at a recent point in time.

With this in mind, we may proceed to what is revealed by the tests.

3.3 Adverbials and PresPerf

The adverbials listed in 3.2 were collocated with the PresPerf of eight predicates: <BE ASLEEP>, <OWN a car>, <SMILE>, <SWITCH OFF the light>, <GROW OLDER>, <KNIT a sweater>, <WIN the race>, which represent most of the event notions that figured prominently in the discussion of PresPerf in 2.1.1. In the following discussion,[13] the results have been augmented by information from a pilot test on another set of predications not listed here.

What seems particularly striking is that a 'resultative' reading is limited to collocation with *for the moment, until further notice*, while in most cases, except, of course, for the 'continuative' reading, which was noted in 1.4 to

13 Throughout this and the following section, I have relied basically on the classification given in Schopf (1984:126-127), and on the more detailed treatment of temporal adverbials presented by Janet Harkness in the 'Tensing System of English' project (cf. her paper in this volume, p.71ff.). I have also benefited from her comments on the earlier version of this paper. Harkness does not explore the question of collocation as such in her analysis of temporal adverbials, so the conclusions I have come to may not necessarily coincide with the terms of her analysis.

require adverbial specification, 'fulfilment' or even 'occurrence' (within a temporal frame) seems to be the general reading. We may take the adverbials in groups whose function is similar. (This does not imply a taxonomy of perfect-collocating adverbials, however.)

3.3.1 Adverbials denoting intervals

3.3.1.1 *For the moment* and *until further notice* collocated with PresPerf assert that at R (= S) a state-of-affairs subsequent to the occurrence of an event holds and will continue to hold into posterior time (after R). They denote, in other words, an interval subsequent to an event. We may depict this, taking an accomplishment as an example, as:

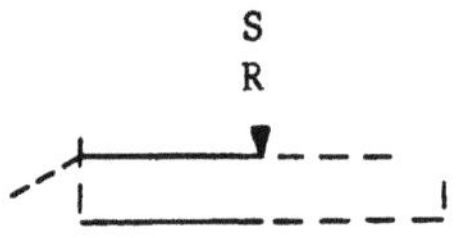

Interestingly, with states and processes which do not have resultant states, the reading is 'fulfilment' or 'accomplishment', i.e. 'X has occurred and will not occur again for a while/until Y'. With event notions involving a resultant state <SWITCH OFF the light>, <LOSE one's purse>, <LOOK AWAY>, <SIT (DOWN) in a chair>, <POUR water into the tank>, <GIVE someone a hammer>, and where the event is reversible, a resultative reading is normal with the implicature that a change or reversal will occur at some stage.

3.3.1.2 Those adverbials that denote, or can denote, an interval up to or including R are those that normally force a 'continuative' interpretation. We may take them in pairs: *all day/all his life, for hours/for years, since 10 o'clock/since he was a boy.*

Only in the case of the *since* + point in time forms is collocation with perfect forms exclusive. With both the *all* + time unit and the *for* + time unit forms, collocation with non-perfect forms is possible, e.g. *He's sitting here all day, He's sitting here for hours, He sits here all day, He sits here for hours.*

In collocation with PresPerf, the *all*-forms will denote an interval that extends, potentially, up to R. This means that a process or state may or may not still be in progress or hold at R. In the case of punctual event notions (punctual event, punctual change and achievement) an 'iterative' (*all day*) or an 'habitual' (*all his life*) reading will be forced. The difference between *all day* and *all his life* pragmatically determines the relative oddity of:

?* He's won the race all day.

compared to the general acceptability of:

He's won the race all his life.

The *for* + time unit type collocated with PresPerf denotes an interval extending up to R. With states and simple processes, the reading is 'continuative'; with punctual changes and punctual events, an 'iterative' or an 'habitual' reading is preferred. Interestingly, PresPerf <KNIT a sweater> and <WIN the race> do not work with *for hours* (i.e. with an 'iterative' reading), but work perfectly well with *for years* (i.e. with an 'habitual' reading).

3.3.2 Adverbials denoting frames or intervals

3.3.2.1 The *since* + time point forms can denote a frame or an interval. The tendency seems to be towards a frame reading at I, especially when the event notion is punctual, but an interval reading at F. Collocation with PresPerf receives a 'fulfilment' reading when *since* + time point sets the frame, but a 'continuative' reading with states and processes and an 'habitual' reading with punctual events, punctual changes, and achievements when *since* + time point is construed as an interval.

3.3.2.2 *Up to now, hitherto*[14] and *so far*. All of these set a frame or an interval which extends up to R. With *up to now*, when it sets a frame, a 'fulfilment' reading is normal, with the possible exception of <OWN a car>, <GROW OLDER> for pragmatic reasons. When *up to now* simply denotes the duration of a state-of-affairs, a 'continuative' reading is favoured by states, quantified processes and incremental processes, an 'habitual' reading by achievements and punctual changes, and an 'iterative' reading by punctual events. With *hitherto*, which typically has a denotation 'on all occasions up to R', an 'habitual' reading is the norm, which for pragmatic reasons seems to exclude <GROW OLDER>. With *so far*, which in addition to setting a

---

14 *Hitherto* is much more readily collocated with PastPerf and PastPerfProg in contemporary usage. When it does occur with PresPerf it is normally in a narrative mode (i.e. in an 'historic present' sequence).

frame or interval, implicates continuation thereof beyond R, PresPerf is consistent with a 'fulfilment' reading (which makes states and non-telic incremental processes aberrant) or 'continuative' (states, non-telic processes), 'habitual' (achievements, accomplishments) or 'iterative' (punctual events).

At this point we should introduce a distinction between intervals which include a number of occurrences of similar events (interval as frame) and those which include a number of occasions on which a similar event occurs, i.e. interval as (sub-)frames. This may be depicted, taking • to represent any event and └┘ to represent occasions (sub-frames) within a frame:

a.

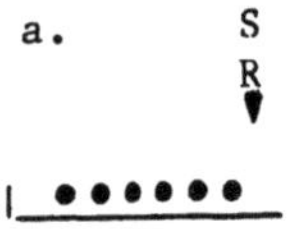

b.

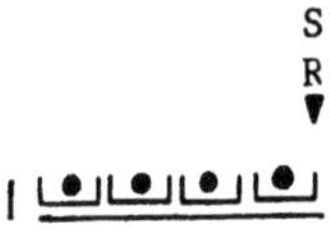

Thus, *Hitherto he's sneezed* is represented better by b, while *Up to now he's sneezed* could be either a or:

For *up to now*, we have the configurations (taking ▬ to represent a state, where a distinction between state and event is necessary):

For *hitherto:*

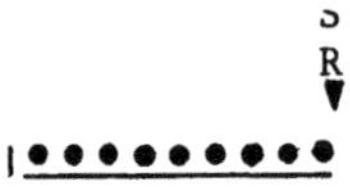

For *so far:*

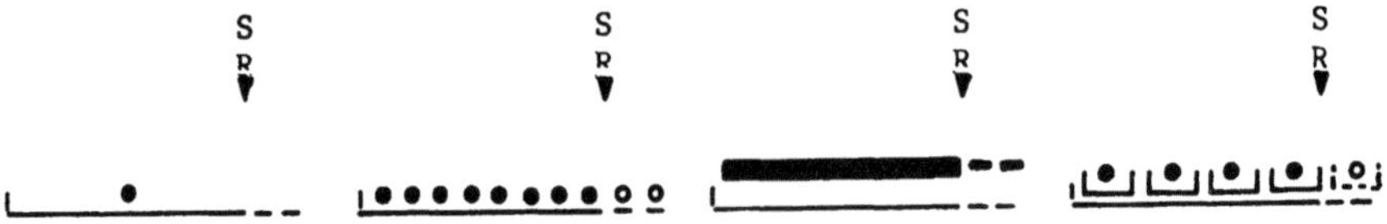

3.3.3 Adverbials denoting frames

3.3.3.1 *Meanwhile*, as a 'setting' adverbial will collocate with temporary states and events, otherwise states do not collocate too well, unless interpretable as behaviour, e.g. *Meanwhile, he's been kind*. It can be interpreted as: 'within temporal frame since last update'. Its configuration might be:

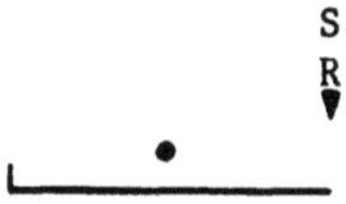

3.3.3.2 With deictically proximal adverbials like *this morning* and *this year*, both of which can establish a 'setting' (at I) or a frame (at F), temporary states are apparently neutral as regards completion at R. With all other event notions a 'fulfilment' reading is the norm. Pragmatic restrictions can be noted with:

He's switched off the light this morning. (fulfilment-resultative)
He's switched off the light this year. (fulfilment-habitual)

and

? He's grown older this morning.
He's grown older this year.

The configuration for *this morning/this year* with PresPerf is as in a below, that for *this morning/this year* with Past as in b.

a.
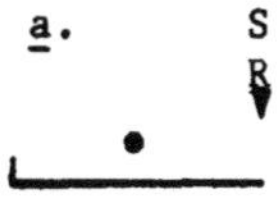

b.

Recently and *lately* are both marked for proximity, but differ from *just* in being capable of 'setting' the frame for an event. Both seem to assert the occurrence or fulfilment of an event within a recent temporal frame extending up to R, but the event or state itself may not hold at R. *Lately* strongly suggests habitual occurrence, except with negative predications, when it seems to be interchangeable with *(not) recently*. (When negative, both involve intervals.) *Lately* may also implicate continuation beyond R. Respective configurations:

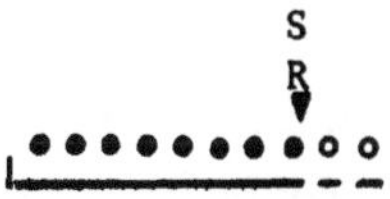

3.3.4 Adverbials implying an anterior frame

3.3.4.1 The function of *now* with PresPerf is primarily to assert fulfilment anterior to R, but exclude any implicature that a change or reversal will occur. As a 'setting' adverbial it will collocate with temporary states and events, otherwise states do not collocate too well, unless interpretable as behaviour, e.g. *He's been kind now,* which is, in fact, an event. It can be interpreted as: 'update on object-world to R'. A configuration for this would be:

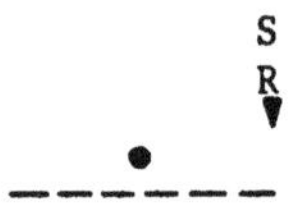

3.3.4.2 *Just* denotes the temporal proximity or recency of occurrence or fulfilment of an event or the recency of the termination of a state. For pragmatic reasons, co-extensive and some non-temporary states do not appear to collocate. *?He's just owned a car.* A configuration for *just* might be:

cf. *He did it just a second ago:*

3.3.4.3 We may turn next to the set of adverbials which, in some way, relate an occurrence (or a projected occurrence) to another occurrence of a (like) event, or to R: *again, before, already, (not) yet.* Here we are concerned simply with occurrence or fulfilment, not with continuation or results. *Again* asserts the repetition of an event or temporary state. It collocates less readily with processes and states that, pragmatically, are unlikely to be repeated. This extends also to events like *die.* With *before,* there is a 'setting' reading: 'on an earlier occasion' and, more commonly, a 'prior occurrence' reading. With the latter, there is a contrast with an actual current or potential (= envisaged) future event/state. With *already*, there is a (rare) 'setting' reading: 'this state-of-affairs has been reached', cf. *And already, a crowd has gathered*, as well as readings which involve implicature such as 'fulfilment proceeds apace - what next?', 'fulfilment established - no need for repetition', and 'premature fulfilment'. With *(not) yet*, expected occurrence or fulfilment is still due. Non-initiated (or non-agentive) predications are less likely to collocate, but by no means excluded. Configurations include:

again

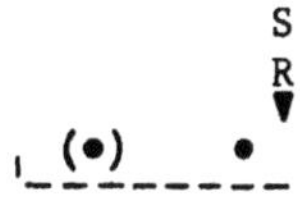

before

already

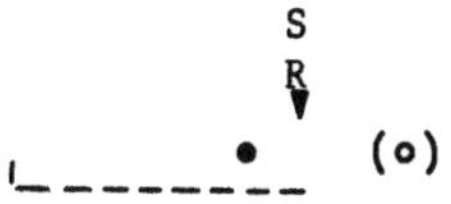

not yet

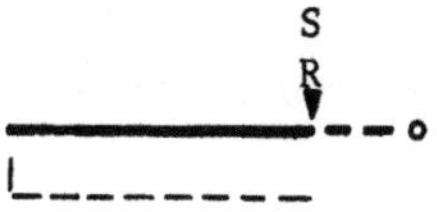

The frame associated with configurations for the perfect appears to be only implicit, i.e. established by virtue of the event relating nature of these adverbials. It may be significant that at least a subset of these (*again, already, not yet*), as well as *just* (3.3.4.2) are precisely those adverbials that differ with respect to collocation with perfect or past tenses between American and British English.

3.3.5 Adverbials of frequency

3.3.5.1 *Once* may give both a 'setting' reading ('on one occasion when I've looked/that I can remember ') or simply quantify the occurrence of the event or state. In the former, pragmatic restrictions will occur with <OWN a car> and <GROW OLDER>, but not with <BE ASLEEP>. A probable configuration for this would be:

With the latter, an obvious pragmatic restriction applies in those cases where an event is not iterable, like <WIN the race>, <KNIT a sweater> unless, of course, the singular nominal is construed as representative for a set (unus pro toto). Here the configuration should be:

3.3.5.2 The reading of *once* + PresPerf comes close to 'experiential', as does that of the pair *never/not (...) ever*. *Never* as 'setting' at I can be glossed 'on no occasion', while *never* and *not ever* at M and *not* at M with *ever* at F is a reinforcement of *not*, i.e. 'not P, and not-P holds for all time'.

3.3.5.3 *Always* can have two readings. When at I as 'setting' it is glossable as 'on all occasions' and is not very likely to collocate with <OWN a car>, <SNEEZE>, <GROW OLDER>, while at M (and contrastively at F) it is glossable either as 'on all occasions' or as 'for all (co-extensive) time'. With the former, the reading of PresPerf + adverbial collocation is 'habitual', with the latter, it is 'continuative', in which case states and (non-telic) processes are favoured and 'once only' telic processes like <KNIT a sweater> are hard to construe, except as 'iterative' or 'habitual', cf. *He´s always sneezed*, where a punctual event <SNEEZE> is construed as a habit, i.e. 'He's always been a sneezer/susceptible to sneezes'. Configurations for *always*:

3.3.5.4 The frequency adverbials *regularly, often, occasionally, usually* and *seldom* can, with the exception of *regularly*, appear at I as 'setting':

?* Regularly, I have washed them. (? contrastive theme)
Often, I have washed them.

When used as 'setting' adverbials, glossable as 'on many/few/most occasions', etc. and collocated with temporary states, even these may be said to be 'eventized' and thus become 'iterative' or 'habitual', as in:

Often, he's been asleep.

i.e. 'on many occasions (when I've called, etc.) he's been asleep, which is not equivalent to: *He's slept often.*

This is pragmatically restricted: <BE TALL> is unlikely, <OWN a car> is possible.[15] With incremental non-telic processes such as <GROW OLDER>, acceptability is questionable. A similar pattern is detectable for the 'frequency of occurrence' reading. Configurations are of the type:

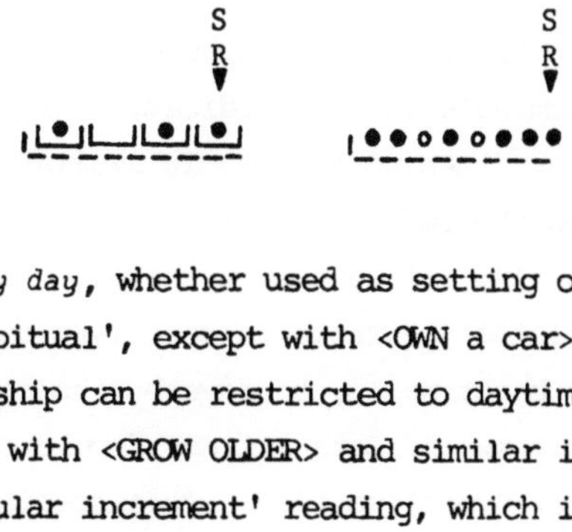

3.3.5.5 With *every day*, whether used as setting or pure iterative adverbial, the reading is 'habitual', except with <OWN a car>, which is pragmatically odd - unless ownership can be restricted to daytime, leaving the nights to non-ownership; and with <GROW OLDER> and similar incremental processes, which allow a 'regular increment' reading, which is, properly, a subtype of 'continuative'. Configurations:

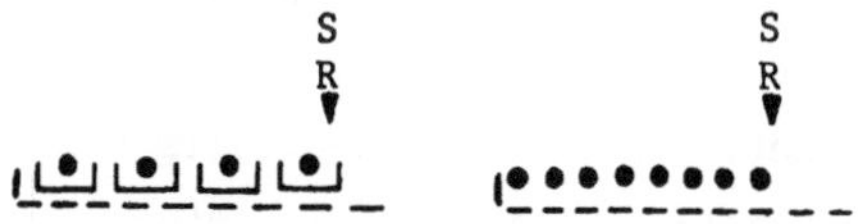

15 *He's owned a car regularly* is something of an oddity in that it does not entail a series of states of owning (different) cars, simply that owning a car (or various cars) has extended over a significant interval, cf. *He's owned a car intermittently (but not regularly enough for us to reduce his insurance rate).*

3.3.6 The remote indefinite adverbial: *in the past*

With *in the past*, an adverbial that was not included in the tests, any reading of a PresPerf is necessarily 'remote indefinite' alias 'experiential'. What is remarkable about this adverbial is that the interval referred to is clearly terminated long before S, but nevertheless the 'current relevance' of the event or state is carried by a the ascriptive nature of the utterance it is contained in and b the current continued existence of the attributee.

3.3.7 Summary

We may complete this section with an initial summary of what has been revealed by the tests.
Clearly, and not surprisingly, adverbials + PresPerf collocations favour a particular reading, which in cases of adverb-dependency and mutual dependency will normally be largely determined by the temporal nature of the adverbial.[16] When the event notion precludes the favoured reading, either an alternative reading presents itself or else the collocation remains uninterpretable.
Thus a 'continuative' reading will be favoured by interval adverbials like *for hours*. If the event notion resists continuation (or better, duration), as is the case with <SNEEZE>, <SWITCH OFF the light>, then an 'iterative' reading (continued state-of-affairs composed of sub-events) will become available. And if the event notion, usually because of the nature of the participants denoted by the nominals in the predication, resists iteration, as is the case with <WIN the race>, then no reading is available.

We should also consider how the reference point (R) relates to the temporal specification given by adverbials. If we accept the proposal of 2.2, whereby the reference point with 'isolated' perfects is conceived of as the point of updating on the state of the object world, an adverb like *now* can be viewed as simply naming this, while *meanwhile* sets a frame from a contextually retrievable t_i (usually another event) up to R. *Just* specifies the proximity of E to R. The relational adverbials such as *already* relate the occurrence of an event via an implicit frame (the state of the world up to R). With adverbials like *for the moment* an interval is established for a state-of-affairs subsequent to the event. This state-of-affairs holds at R. An adverbial like *since 10 o'clock* establishes the t_i of an interval or frame which

16 In the case of adverb-dependency, one would expect complete determination. But the adverbial itself may be ambiguous.

extends up to R; others such as *so far, for hours* do not have specified t_i; and at least in some cases *so far* may involve a frame that is not concluded at R. In the case of frequency adverbials, the events are related via the implicit frame that includes them to R. In the case of *every day* the (non-specific) points of the events (E) are given (e.g. *He's been there every day*) or else sub-frames are set up (e.g. *Every day, he's gone there, sat with her, fed her, and nursed her ...*). In both cases, these events or sub-frames are related via a frame extending up to R.

Finally, we may note that examples can be found where it appears that a specific time is given by an adverbial collocating with PresPerf:[17]

He has been at work at 6 a.m. (, but it wasn't a very productive day).

Unlike *at 6 a.m.* collocated with Past or PastProg, where it can be maintained that the adverbial is the specification of R, we have here the denotation of a set of times within a frame up to R, on at least one occurrence of which the state in question <BE at work> held. It does not, however, directly specify the point of the event (E) in Reichenbachian (or even Schopfian) terms.

3.4 Adverbials and PresPerfProg

The same set of adverbials was collocated with the PresPerfProg of eight predicates, seven of them identical to those in 3.3, <BE ASLEEP>, however, being replaced by <LIVE in Paris> because of the former's non-progressivizability. Although <OWN a car> is often considered to resist progressivization, e.g. **I am owning a car*, there are a number of acceptable progressive forms: *In a week's time, if we're lucky, we may be owning a car. All the time they were owning a car, they never once used it on a Sunday.* It was kept, therefore, in order to see whether collocation with a particular set of adverbials improved its acceptability.

The results of these tests will be discussed below, without, however, going into quite so much detail as with PresPerf and adverbials. It will be assumed, unless explicitly claimed to the contrary, that the readings of the adverbials remain constant.

---

17 I was reminded of this possibility by John Lyons (in conversation).

3.4.1 Adverbials denoting intervals

3.4.1.1 With *until further notice* collocation does not seem to work even in cases like <BE SWITCHING OFF the light>, where with PresPerf a clear resultative reading was possible. There may, however, be an interpretation with *until further notice* as part of the predicate: *What's he been up to? He's been living in Paris until further notice, just as you told him.*

With *for the moment*, however, the collocation appears to work with <BE KNITTING a sweater>, <BE SWITCHING OFF the light> and other predicates when they have an 'agentive' subject. It is important to note that PresPerfProg merely selects a phase of the event, e.g. the phase leading up to achievement (incremental process) in the case of <BE WINNING the race> or, in the case of <BE SWITCHING OFF the light>, a phase of seriality (repeated occurrences). With <BE LIVING in Paris> a 'temporary habit' reading seems possible. In none of these cases do we have a 'resultative' or 'concomitant overlapping' reading.

3.4.1.2 With intervals up to and including R (= S), the pattern of meaning is clear. States, simple, quantified and incremental processes and accomplishments will favour a 'continuative' reading, e.g.

He's been knitting a sweater all day/for hours/since 10 o'clock.

while punctual events and changes favour 'iterative' (i.e. serial activity) readings, e.g.

He's been sneezing all day/for hours/since 10 o'clock.

With an achievement like <WIN the race> a 'continuative' reading is possible. (continuation of the lead-up phase) but presupposes in pragmatic terms a longish race, cf. *He's been winning the race all day.* With *for years, all his life, since he was a boy*, pragmatic factors will favour an 'habitual' reading. This is also the case with other event notions, cf.

He's been switching off the light all day.
He's been switching off the light all his life.

where we most naturally interpret the former as repeated actions and the latter as habitual action. *He's been switching off the light for hours* could be 'iterative' (serial activity) or 'conative', i.e. activity in the phase prior to the mutation or transition phase.

3.4.2 Adverbials denoting intervals or frames

3.4.2.1 The *since* + time point adverbial does not appear to allow a frame interpretation when collocated with PresPerfProg, e.g. in *Since 10 o'clock, I've been feeding the chickens.*

3.4.2.2 With *up to now* and *so far*, we find 'continuative' readings for states and processes, 'serial acitivity' for punctual events, 'activity' and 'habitual' for changes. *Up to now* as an interval favours 'continuative' readings, *so far*, which may be a frame or interval, favours 'activity' or 'incremental process' readings, except when the event notion allows (or even forces) a 'continuative' reading. There is, therefore, a potential distinction between:

Up to now, she's been knitting a sweater.
So far, she's been knitting a sweater.

with *up to now* denoting an interval over which <she BE KNITTING a sweater> holds, and *so far* denoting a frame within which <she BE KNITTING a sweater> has taken place.

Hitherto collocates more successfully with <BE LIVING in Paris> and other non-punctual predicates to give 'temporary habit' or 'limited state' readings than with punctual predicates to give 'serial activity'. Its preferred position is at I. This seems to support the view that it denotes an interval of sub-frames.

3.4.3 Adverbials denoting frames

3.4.3.1 With *meanwhile*, PresPerfProg gives us generally an activity which may or may not have proceeded to its natural conclusion: *Meanwhile, he's been winning the race* is genuinely indeterminate as to whether the race has been won or not. <BE SWITCHING OFF the light> also permits an 'iterative' (i.e. serial activity) reading. <BE OWNING a car> does not work at all. With <BE LIVING in Paris> we get a 'temporary habit' or 'limited state' reading.

3.4.3.2 With *this morning* and *this year*, states give a 'temporary habit' or 'limited state' reading, e.g. *He's been living in Paris this year*, though for pragmatic reasons *this morning* would scarcely work. The state-of-affairs may, however, not extend up to R, cf. interval adverbials.

> Although he's been living in Paris this year, he's now living in Toulouse.
>
> * Although he's been living in Paris since he was a boy, he's now living in Toulouse.

Otherwise 'punctuals' will favour an 'iterative' reading, while 'non-punctuals' favour an 'activity' or 'incremental process' reading, with *this morning*, sometimes becoming tantamount to 'temporary habitual' readings with *this year*. Cf. *He's been smiling this morning* (activity within the frame of *this morning*, possibly, but not necessarily co-extensive with *this morning*) and *He's been smiling this year* (activity within the frame of *this year*: presumably recurrent, but not necessarily).

3.4.3.3 With *recently* and *lately*, the difference we observed in 3.3.3.3 can be discerned again: *lately* tends to favour an habitual reading (even if it is only a temporary habit) and imply continuation of the interval up to R (= S), while *recently* favours activities or series of events within a frame up to R (= S). Thus *he's been living in Paris recently* suggests a recent temporary state-of-affairs which may or may not hold at R (= S), while *he's been living in Paris lately* suggests that the referent of *he* has 'taken to' living in Paris, that his current wont is to be resident in Paris despite potential occasional absences.

3.4.4 Adverbials implying an anterior frame

3.4.4.1 With *now* + PresPerfProg collocations we generally find the readings associated with isolated PresPerfProg: 'temporary habit' for <BE LIVING in Paris>, 'activity' for <BE SMILING>, <BE KNITTING a sweater>, 'serial activity' for <BE SNEEZING>, 'activity in pre-phase' or 'serial activity' for <BE SWITCHING OFF the light>. <BE OWNING a car> does not collocate well, except with a sense change with *now* at I. With <BE GROWING OLDER> and <BE WINNING the race> an 'incremental process' reading, contrary to our expectations perhaps, is only marginally possible. An 'activity' reading for <BE WINNING the race> does not seem possible.

3.4.4.2 With the probable exception of <BE OWNING a car>, *just* collocates readily with PresPerfProg and often implicates the completion of an activity: *He's just been switching off the light*. <BE WINNING the race> collocated with *just* gives us an activity (what X has been up to) or the state of the race: *He's just been winning the race, but it looks as if Thunderstruck may yet edge ahead of him.*

3.4.4.3 As might be expected from the discussion in 3.3.4.3, the readings we have with relational adverbials involve the occurrence of an activity or process, or the obtaining of a state-of-affairs (in the case of *again, before, already*), or the non-occurrence or non-obtaining (in the case of *not yet*) within a frame extending up to R (= S) that is implicitly set up by the adverbials. Thus: *He's been smiling before* asserts that the state-of-affairs (the process phase of <SMILE>) has held within an implicit temporal frame extending up to R. And: *He hasn't been smiling yet* asserts that the (expected) state-of-affairs has not held at any time within the interval to date, i.e. R.

3.4.5 Frequency Adverbials

3.4.5.1 There is a severe restriction on the collocatability of *once* with PresPerfProg. *Once* favours an 'occurrence (within frame)' reading, which in turn favours an event rather than a state-of-affairs. Thus *once* in *He's been knitting a sweater once* can only be construed as 'on one occasion'. This is even more evident at I, e.g. *Once, he's been smiling*, which can be read as: 'once (when I've looked at him/in on him, etc.), he's been smiling'.

3.4.5.2 With *never/not ever*, the collocation is unproblematic and can be read as 'on no occasion (within a frame up to R) has the state-of-affairs obtained.

3.4.5.3 With *always* we find either an habitual (or temporary habitual) reading, mainly with non-states, or else a 'continuative' reading, mainly with states or non-quantified processes. *Always, he's been smiling* can mean 'on all occasions when (X)' ('setting' function and habitual reading) or, less obviously, 'for all time' ('contrastive thematic' function and 'continuative' reading), while *he's always been smiling* would be more likely to have a 'continuative' reading.

It becomes clear that <BE OWNING a car> collocates readily with *never/not ... ever* and *always* as 'setting' adverbials.

3.4.5.4 With *regularly* collocation is restricted: the adverbial normally only occurs at M or F, and the reading is habitual, except perhaps in the case of <SNEEZE>, which may be interpreted as instances of sneezes at regular intervals.

3.4.5.5 The remaining frequency adverbials *often, occasionally, usually* and *seldom* readily provide settings for states-of-affairs. Thus: *Often, he's been sneezing* is normally to be read as: 'on frequent occasions (when I've been there, etc.) he's been undergoing a bout of sneezing (= a series of sneezes)'. A less obvious reading with adverbial at F, e.g. *He's been sneezing often* is a quantification of the subevents, i.e. the number of sneezes. A 'contrastive rheme' at F may imply some kind of interruption.

> He's been winning the race often, but his legs have always given way in the last furlong.

cf.

> Often he's been winning the race when his legs have given way in the last furlong.

3.4.5.6 With *every day*, the usual reading is 'iterative' (serial activity) or 'habitual' activity with the implicature in cases like <BE WINNING the race> of fulfilment, cf. *He's been winning the race every day.*

3.4.6 The remote indefinite adverbial: *in the past*

In the past can only collocate, apparently, with predications that are iterable and can, when progressivized denote a process phase or serial activity. The reading is that of a remote temporary habit, as in: *In the past he's been smiling.* It denotes an interval of occasions when the activity or state-of-affairs has obtained.

3.4.7 Summary

We may conclude this examination with a brief resumé of the kind of readings that have been found with adverbial + PresPerfProg collocations. Typically, a progressive form denotes some kind of state-of-affairs (whichever phase the progressive selects from the event notion). This can be set in the "anterior time zone" in a number of ways. The simplest is the one that is most like the 'isolated' PresPerfProg. Here the state-of-affairs (activity, incremental process, serial activity or (temporary) state) is contained within the frame with no further limitation, i.e.

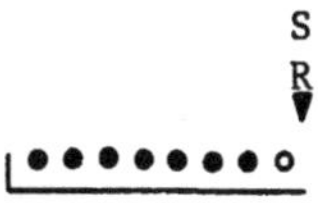

cf. singing, growing wiser, sneezing, hating her, lying on the floor
  + *this morning, recently*

A second is where the activity, etc. is within the frame but excluded from simultaneity with R:

cf. singing, growing wiser, sneezing, lying on the floor
  + *now, meanwhile*

and similarly, except that an actual or envisaged second activity is involved, with *before, already, again.*

A third is where the activity, etc. is proximate to R but necessarily prior to it:

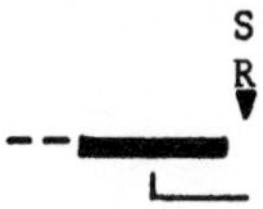

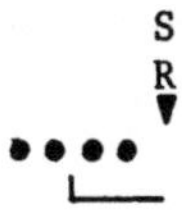

cf. singing, expanding, sneezing, lying on the floor
  + *just*

A fourth is where the activity, etc. continues co-extensively with the frame (which then becomes the interval for the activity):

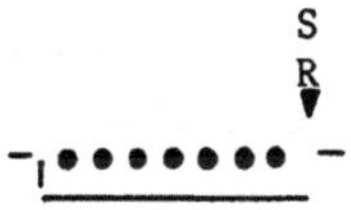

cf. singing, increasing, sneezing, lying on the floor
  + *all day, for hours, since 10 o'clock*

And a fifth is where the activity, etc. takes place at one or more points (occasions) within the frame:

cf. singing, growing, sneezing, lying on the floor
+ *often, occasionally, every day* (at I)

This differs somewhat from the PresPerf, where the events are within sub-frames,

cf.

Occasionally, he's sung.

Occasionally, he's been singing.

This configuration is also what we get with *every day* at F, e.g. *He's been singing every day.*

4 Evaluation and interpretation of the results

We should now attempt to extrapolate from the somewhat jumbled facts in 3.3 and 3.4 a set of general hypotheses about how the collocation of perfect tenses and the adverbials examined will be interpreted semantically. It will be useful to reformulate and recapitulate first the hypotheses of 2.1.1 and 2.1.2 on the 'isolated' perfect.

4.1 'Isolated' Perfects

4.1.1 PresPerf

PresPerf locates the whole event notion associated with a predicate within an anterior temporal interval ("anterior time zone" or frame) extending up to R (= S).

> If the event notion is non-dynamic, it will be understood as being terminated within the anterior temporal interval.
>
> > The event/state will be proximal if the event notion is temporary, distal if it is non-temporary.
>
> If the event notion is dynamic, it will be terminated (if non-conclusive), or completed if conclusive within the anterior temporal interval.
>
> > If it is agentive, there will be an implicature of achievement or fulfilment; and if it is incremental or transitional and conclusive, there will be an implicature that the resultant state obtains at R (= S).
>
> All event notions, except those that are non-iterable and non-reversible, will, under suitable discourse conditions (ascription), be interpretable as 'experiential'.

The class of conclusives, in Schopfian terms, consists of quantified processes, punctual events, punctual changes, quantified incremental processes, and accomplishments.

The class of incrementals consists of non-quantified incremental processes, initiated incremental processes, quantified incremental processes, and accomplishments.

The class of transitionals consists of punctual changes and achievements.

The above is not intended to explain the case of <KNOW (fact, etc.)>, which is notorious in resisting PresPerf, except with an interval adverbial (and also resisting PresProg, and thus PresPerfProg), or the perfecto-praesentium *HAVE got*.

4.1.2 PresPerfProg

It will be useful to recall what groupings event notions may be divided into in terms of their progressivizability.

> A. Temporary states <BE asleep> etc. and temporary physical conditions <BE red in the face> etc., relational <CONTAIN drugs> etc. and mental states <BELIEVE that + Clause> etc. do not normally progressivize.

(Those that do will have an actional reading (i.e. current behaviour).)

B. Co-extensive/durative states <BE tall> etc. and qualities <BE friendly> etc. may progressivize if they can take an actional meaning (i.e. denote current behaviour).

C. Emotional states <LOVE her/the play> etc., perceptual states <SEE her> etc., bodily sensations <(My feet) HURT> etc., potential <RULE England> etc. and habitual <DRIVE a cab> predications progressivize in taking on a temporary sense.

(N.B. some perceptual states may become actional, in which case they may follow set E.)

D. Locative manner <LIE on the floor>, simple processes <BREATHE>, <RAIN> etc., initiated processes <SING> and quantified processes <LAUGH> and punctual changes (type 2) <LOOK AWAY> etc. progressivize in selecting only the process phase.

E. Punctual events <KNOCK> etc. and punctual changes (type 1) <SWITCH OFF the light> etc. progressivize in denoting a process of iteration (series of sub-events).
Punctual changes (type 1) can also select a prior phase (conation or inception phase).

F. Punctual changes (type 3) <BEGIN to rain> etc., non-quantified incremental processes <(The universe) EXPAND> etc., initiated incremental processes <IMPROVE> etc., quantified incremental processes <(The hay) DRY> and accomplishments <BUILD a house>, <CLEAN the room> etc. select an incremental process phase.

G. Achievements when progressive may select either the incremental process phase they imply <DEFEAT the enemy> etc. or the phase immediately prior to achievement <REACH the station> or set up a process of iteration (with plural subjects, etc.) <(The guests) ARRIVE>.

PresPerfProg locates that phase of the event notion associated with a predicate that is selected by progressive aspect taking account of any sense change (cf. above and Schopf (1984:235-257)) within an anterior temporal interval extending up to R (= S).

If the event notion is non-dynamic, the implicature will be that of a temporary state-of-affairs.

If the event notion is dynamic, then the phase selected will be understood as an activity, serial in the case of punctual event notions, incremental in the case of incremental event notions, and so on.

If the event notion is one which permits selection of a prior phase, there will be a strong implication (but not assertion) of completion of phase.

The only apparent exception to this is the behaviour of relational event notions, some of which, though not normally appearing in PresProg, can, for some speakers, be cast as PresPerfProg with a limited state interpretation. This is probably to be explained as a formation analogical to potential predications like <RULE England>.

4.2 'Collocated' Perfects

Without analyzing in detail the adverbials tested, we may predict on the basis of their denotation the kind of perfect reading that will emerge from collocation. Adverbials that may function in more than one way temporally will, according to this hypothesis, be in principle consistent with different readings of perfect + adverbial collocation.

4.2.1 Adverbial + PresPerf collocations

If an adverbial denotes an interval including and extending beyond R (= S) (e.g. *until further notice, for the moment*) then

a non-conclusive and iterable event notion will be understood as having been fulfilled or achieved prior to the interval.
a conclusive and reversible event notion will be understood as having been completed prior to the interval with the resultant phase still obtaining.

(This appears to be subject to considerable pragmatic restriction.)

If an adverbial denotes (or may denote) an interval extending up to R (= S) or extending to include R (= S) (e.g. *all day/his life*, etc.) then

a non-punctual event notion will be understood as continuative, i.e. concurrent with the interval.
a punctual event notion will be understood as iterative (provided it is iterable), i.e. repeated throughout the interval.

If, in addition, an interval is, in empirical terms, long relative to the normal duration of the event, then a dynamic event notion (punctual or non-punctual) will - provided it is iterable - be understood as habitual.

If an adverbial denotes an encompassing interval (frame) (e.g. *meanwhile*), then

a dynamic and/or temporary event notion will be understood as having occurred within the frame.
a non-dynamic and non-temporary event notion will be understood as obtaining for a limited period within the frame.

If an adverbial denotes an interval of occasions (e.g. *lately*), then these will be understood as sub-frames, within which any event notion will be understood as having occurred

i.e. habitual occurrence if dynamic and/or temporary.
i.e. habitual state if non-dynamic and non-temporary.

If an adverbial does not denote an interval but only implies one by virtue of relating an event or state with R (= S) (e.g. *now; just*) or occurrence with another actual or potential occurrence assessed at R (= S) (e.g. *before; again*), then

any event notion will be understood simply as having occurred (co-extensive states and non-iterable events and states will be subject to restriction).

If an adverbial denotes frequency (e.g. *often*), then -provided the event or state is iterable -

the event notion will be understood as occurring with the relevant frequency.

If an adverbial denotes regularity (e.g. *regularly*), then - provided the event or state is iterable -

the event notion will be understood as occurring with the relevant regularity.

If an adverbial denotes frequency and regularity (e.g. *every day*), then - provided the event or state is iterable -

a non-conclusive incremental event notion will be understood as a continual increment.
a conclusive or non-incremental event notion will be understood as iterative, or depending on the event, habitual.

4.2.2 Adverbial + PresPerfProg collocations

The classification of the semantic effects of progressivization inventarized in 4.1.2 applies equally here, and it will not be referred to explicitly in what follows.

If an adverbial denotes an interval which includes and extends beyond R (= S) (e.g. *for the moment*), then subject to considerable non-obvious restriction, an event notion cast as a PresPerfProg predication will be understood as obtaining (at least potentially) in the phase selected by the progressive throughout this interval up to R (= S), i.e. 'temporary activity'.

(There is, therefore, a significant distinction between these adverbials when collocated with PresPerfProg and when collocated with PresPerf.)

If an adverbial denotes an interval extending up to R (= S) or extending to include R (= S) (e.g. *all day*, etc.) then the phase of the event notion selected by progressivization will be understood to extend throughout that interval, i.e. the reading will be 'continuative': continuation of limited state, activity, serial activity, inceptive activity, etc.

If such an interval is, in empirical terms, long relative to the normal duration of the event, then - provided the event notion is dynamic - a 'temporally limited habit' reading will be usual.

If an adverbial denotes an encompassing interval (frame), then the phase of the event notion selected by progressivization will be understood as having occurred within that interval.

If an adverbial denotes an interval of occasions, then these will be understood as time-points at which the phase of the event notion selected by the progressive will have obtained.

If an adverbial does not denote an interval but relates an event or state to R (= S) or to another actual or potential occurrence assessed at R (= S), thus implying a frame, then the phase of the event selected by progressivization will be understood to have obtained within the implicit frame.

If an adverbial denotes frequency, regularity or frequency and regularity (e.g. *often; regularly; every day* respectively), then - provided the event notion is iterable - the phase selected by the progressive will be understood as obtaining with the relevant frequency and/or regularity.

4.3 Conclusion

We have, in the course of this paper, presented a number of findings on individual predicates and collocations with adverbials. Because of their detailed nature, it would be nigh on impossible to recapitulate them here. The general conclusions we have come to, however, may quite profitably be re-stated.

In §1, we argued that the polysemic view of the Present Perfect, involving as it does numerous factors that are pragmatic in the broadest sense - presuppositions and implicatures, communicative function, descriptive mode etc. - cannot be upheld theoretically. Though it looked as if the terms of most polysemic accounts could be derived from and subsumed under the single term of a monosemic account (whatever its nature), we did not commit ourselves at that point.

In §2 it was argued that PresPerf and PresPerfProg, in isolation, should be viewed as cases where the event notion as a whole or the phase selected by progressive aspect, respectively, is pushed into the "anterior time zone". Since the readings ascribed to PresPerf and PresPerfProg turn out either to be chimerical (e.g. 'hot news') or else accountable for as consequences of the event notion (e.g. 'resultative') we can adopt the view that the 'isolated' PresPerf and PresPerfProg are 'monosemic' in having the kind of configuration just described. This "anterior time zone" consists minimally of a frame extending from an unspecified anterior point in time up to R (the "stock taking" reference point located at S).

In §3 we illustrated how this basic configuration may be modified or made more specific when PresPerf and PresPerfProg are collocated with adverbials. In many cases the adverbial is determinative of the reading. A number of variations in reading can be explained on the basis of event notion, but again object-world pragmatics must sometimes be appealed to. In particular, notions such as agentivity, iterability, reversability, etc. have proved relevant. These may be subsumed under 'causality' - a parameter in the meaning of event notions that needs exploration.

The results of the tests have been formulated as a set of hypotheses above (4.1 and 4.2). These seemed to be the stuff of which systemic analyses are made, and a modest attempt at this has been undertaken in Tables VII a-d (p.173ff.). Though fragmentary (redundancy, for example, has not been eliminated, nor has integration of the system fragments been attempted), they may be taken as a step towards specifying what Sørensen terms the "basic unity" of the Perfect.

Table VII(a) System fragment for interpretation of isolated Present Perfect

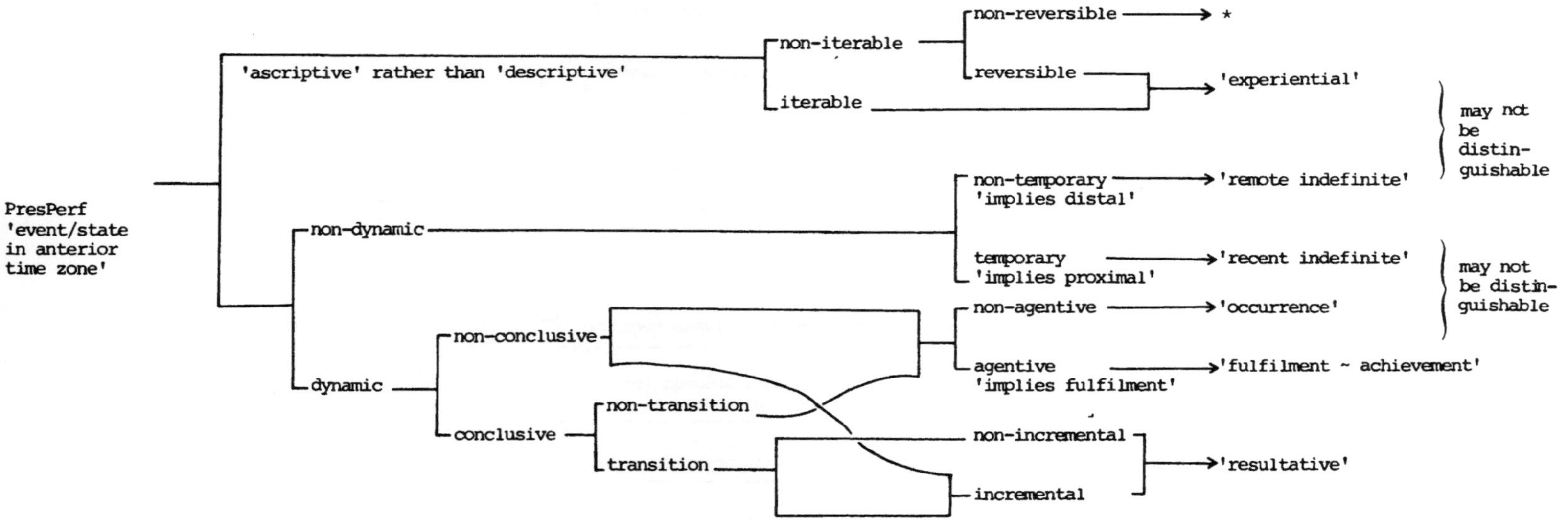

Table VII(b) System fragment for interpretation of isolated Present Perfect Progressive

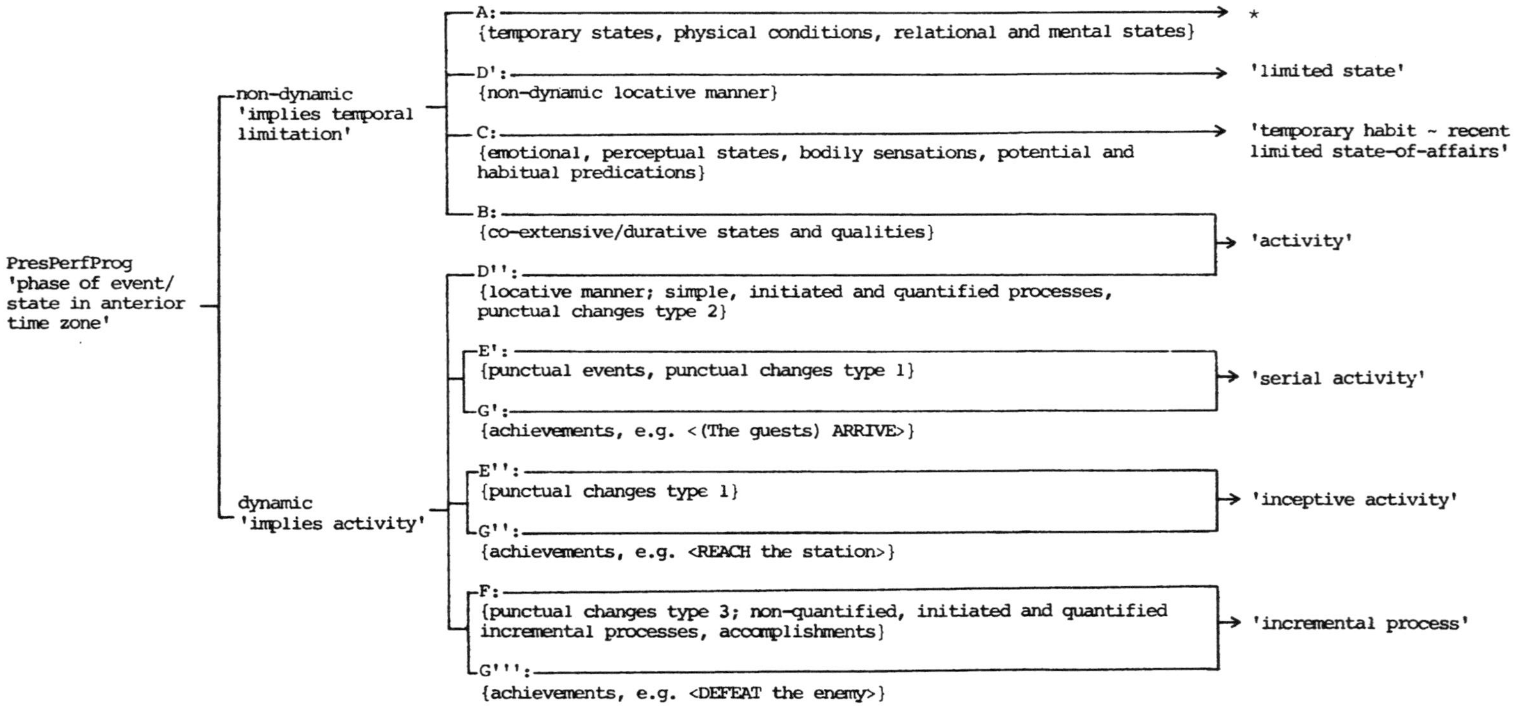

Table VII(c) System fragment for interpretation of Adverbial + PresPerf collocations

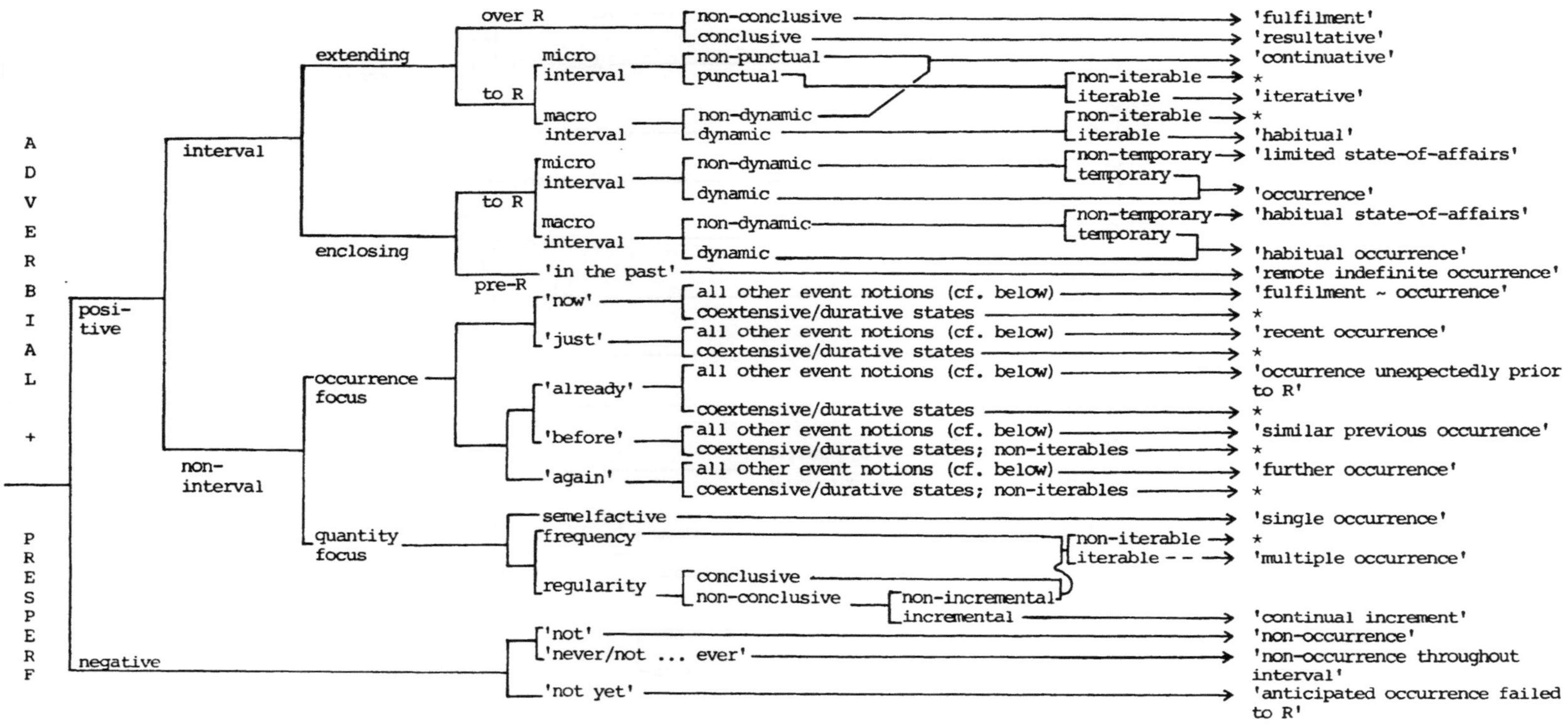

Table VII(d) System Fragment for interpretation of Adverbial + PresPerfProg collocations

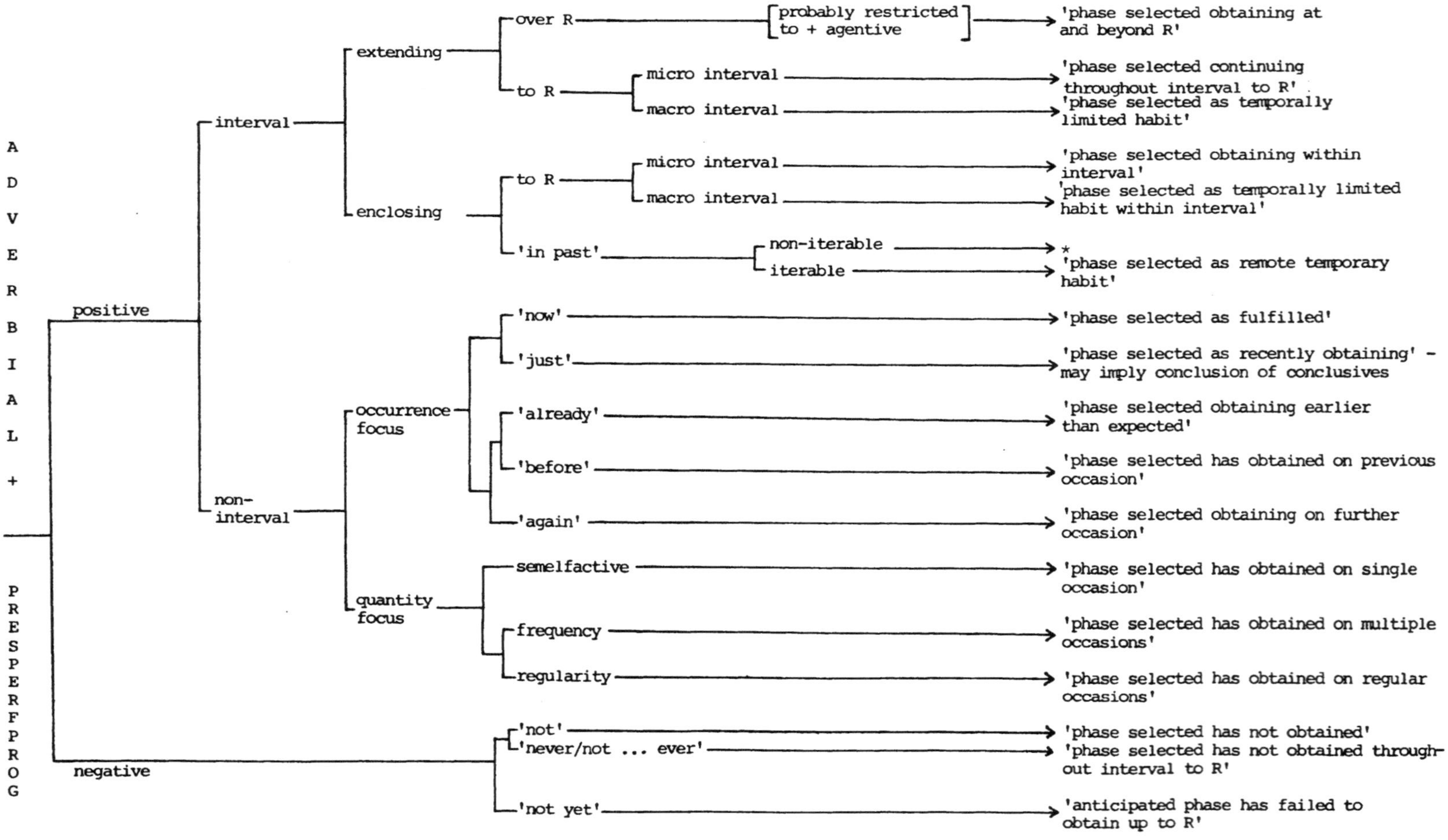

THE PAST TENSE IN ENGLISH

Alfred Schopf

1.0 Past Tense and Present Perfect Tense as compared in traditional studies

Studies of the Past tense in English have in the past almost always been comparative in that, for example, the uses and functions of the Past tense have been compared with those of the Present Perfect or Present tense. The observational data that were considered most revealing in this respect were the collocability of the various tenses with the adverbials. The study of these regularities very soon revealed that the domain of the Past tense is the true past, i.e., does not include the moment of speech. The domain of the Present Perfect, in contrast, is also that of the past but includes the moment of speech. Early studies expressed this idea by saying that the past time referred to by the Past tense must be separated from the moment of speaking by a time gap, i.e. must be a "definite" past time with boundaries in the past. The time involved in the Present Perfect on the other hand was an "indefinite" past time, i.e., had no right-hand boundaries and, therefore, included speech time.

1.1 The Present Perfect- a retrospective Present: O. Jespersen

Otto Jespersen developed earlier ideas along these lines still further by saying that the Past (in his terminology: the Preterit) "refers to some time in the past without telling anything about the connexion with the present moment, while the Perfect is a retrospective present, which connects a past occurrence with the present time, either as continued up to the present moment or as having results or consequences bearing on the present moment."[1]

1.2 The Past Tense separated from the present by a time gap

Linguists concluded from this that an event reported in the Past tense is looked at from a point of view in the past, whereas to report an event in the Present Perfect tense involves looking at it from the point of view of the moment of speech. Another way of putting it was to say that an event reported in the

1 Jespersen (1933) and (1909-1949).

Present Perfect is considered to have "current" relevance, whereas the Past tense carries the information (perhaps as an implicature only) that the situation reported no longer obtained, i.e., was no longer true at speech-time. This idea has been expressed by several linguists including Gerhard Dietrich[2] and Bernard Comrie[3].

1.3 Friedrich Brinkmann and the Present Perfect

A more important observation had been made by Friedrich Brinkmann several decades before. In characterizing the Present Perfect he said that this tense never relates the individual events reported to one another either as sequential or simultaneous.[4] This is undoubtedly an important observation, especially if we contrast it with the observation that several events reported in the Past tense normally tend to be interpreted as simultaneous or sequential and only marginally allow an interpretation which takes them as temporally unrelated.

1.4 Hans Reichenbach's analysis of the Past and Present Perfect in English

Brinkmann's observations, although undoubtedly correct, were hardly taken notice of. They did, however, find an explanation in Hans Reichenbach's trichotomic analysis of the English tenses.[5] In his analysis, the Past tense relates an event (E) as simultaneous to a specific reference time (R) anterior to speech time: $\underset{E}{R}\longrightarrow\overset{S}{}$. The Present Perfect, however, reports an event as having occurred anterior to a reference time which coincides with speech time: $\overset{E}{}\longrightarrow\underset{S}{R}$. This then incorporates and explains the observations made by Brinkmann: If we assume that every event narrated in the Past tense has to be related to a reference time, a group of events will necessarily either be related to one and the same reference point or each of these events will have its own reference time such that the events will be sequentially ordered. The Present Perfect, however, does _not_ relate the event reported to a reference time in the past. It simply says that it occurred at some time anterior to speech time. The two values so far identified for the Past and Present Perfect tenses were given unusual names by W.E. Bull.[6] He distinguishes between _vectors_, _scalars_ and _tensors_. These terms, however,

2 Dietrich (1955:144).
3 Comrie (1985:24).
4 Brinkmann (1906:719ff.).
5 Reichenbach (1947:287ff.).
6 Bull (1968).

should not be understood in their mathematical sense. Vectors in Bull's sense simply state whether a time is earlier, later or the same time as another time. Scalars name the numerical value of a temporal extension or distance in time measured in time units: two days, several weeks etc. The combination of a vectorial value or temporal direction and scalar or numerical value of an extension yields Bull's tensor. Bull seems to think that the description of the English tenses must be based on the following three factors: (1) four orientation axes or anchors: Present, Past, Future and Future in the Past; (2) three vectorial values: simultaneity, posteriority and anteriority; and (3) tensorial specifications. The Past tense in English would in Bull's approach probably have to be interpreted as a zero-tensor (or a simultaneity vector) anchored to a past orientation axis.

1.5 Tense logic and the Present Perfect

Stimulating contributions to the analysis of the tenses have also come from tense logic:

1.5.1 N. Rescher and A. Urquart: The Past Tense as a Present Perfect?

Rescher and Urquart[7] have proposed an indefinite interpretation of the tenses, i.e. of the Past and the Future tenses. The Past tense is paraphrased as "It has been that p" (where p stands for the timeless proposition) or as "p has been realized at some time in the past."[8] The paraphrase in the Present Perfect indicates that the Past tense is attributed a vectorial value only. In the formula proposed for the Past tense, viz. $Pp = (\exists t)\left[U\ t\ n \& R_t(p)\right]$, the vectorial value is expressed by the symbol denoting temporal precedence, U. The formula uses the following further symbols: n = now: t = a time, p = the proposition, and R_t (p) = "the proposition realized at t." The latter should in my opinion be paraphrased as "p simultaneous to t". The existential quantifier, $\exists t$ = "there is at least one time".

The formula clearly shows that the Past tense contains two temporal relations: a relation of temporal precedence U (t,n): a time (t) earlier than now (n); and a relation of simultaneity: R_t (p) = "the proposition (p) realized at, or simultaneous with, t. A sentence in the Past tense, e.g. *The light went out*, would therefore have to be paraphrased as "There is at

---

7 Rescher/Urquart (1971).
8 ibid. (1971:52).

least one time in the past at which the light went out" and such a sentence would be true at speech time if there were any time at all in the past at which p. But this is apparently not the information our sentence is intended to convey.

1.5.2 B. Partee and the definite interpretation of the Past tense

Barbara Partee therefore proposes a definite interpretation of sentences in the Past tense: The tense in *I didn't turn off the stove* is, according to her analysis, used to refer to an understood particular time in the past.[9]

Another important suggestion Barbara Partee makes is the proposal to look at propositions as being true at time intervals rather than simply at time points. She seems to endorse the intuitively correct notion that a speech act occurs at a moment in time and that, therefore, a statement can be looked upon as being true at the moment the assertive function of a statement gains force.[10] But she finds it counter-intuitive that a sentence such as *John builds a house* should be looked upon as true at a moment of time, e.g. "at the moment when he drives in the last nail."[11] In order to achieve as much generality as possible she proposes taking moments of time as (the limiting case of) intervals, which allows us to consider all sentences to be true at intervals.

This, however, raises the question of at which intervals propositions representing various event notions can be considered to be true. There is, for example, no doubt that states can be true at moments (points) of time, in as much as we can relate a state to a moment of time as encompassing this moment: *At 10 o'clock the house was on fire*. We cannot, however, relate an accomplishment in this way to a moment of time:* *He built the house at 10 o'clock*. We shall return to consider this question at a later point.

A third question raised by Barbara Partee is whether tenses, particularly the Past tense, should not be looked upon as instances of temporal anaphora. In other words, the Past tense should not be understood as meaning "at some time in the past," but as referring to some definite past time, the specification of which is provided by a non-linguistic or linguistic antecedent."[12] We consider this an illuminating suggestion. However, we interpret the Past tense not simply as anaphoric temporal reference, but as a search instruction, as will be indicated below.

---

9 Partee (1973:601-609).
10 Bennett/Partee (1978:10).
11 ibid. (1978:11).
12 Partee (1984:245).

2.0 Event notions and reference time

To temporalize a proposition is to relate its event to the time line via a reference point or reference time. We should not overlook the fact, however, that the various event notions as realized by timeless propositions require different reference times. "Event notion", as used here, can be understood as the phasal structuring of events, "event" being used in a broad sense to include, for example, states and processes.

2.1 How to relate states to the reference time

States extend in time homogeneously. They have no inherent initiation phase or end-point. To locate a state on the time line, a reference time is required which selects a "sub-interval" of the state, in many cases a time-point. As states lack an inherent initiation and termination phase, they require internal temporalization, i.e. whenever we relate a state to a time point or time span, the state must necessarily surround, include or overlap it. They can never be temporalized initially or finally. The lack of these phases manifests itself also in the time-comparing sentence (the coincidence formula) or in the "Inzidenzschema" (to use German terminology). This sentence type consists of a clause functioning as a time adverbial and a main clause signalling an event: *When John came home, Mary was drunk.* This sentence denotes a situation in which the state of drunkenness surrounds or overlaps with John's arrival at home.

In some cases, however, the state denoted by the main clause does not overlap with the "when-event" but is sequential to it: *When I turned off the light, it was pitchdark around me.* This might seem to invalidate the principle that states surround their reference time, but since the state named in the main clause (=darkness) is the resultant state of the "when-event" - which means that their successive ordering can be traced back to information supplied in the text - we can uphold our principle.

State propositions can of course be combined with adverbials indicating duration: *We waited for three hours.* These adverbials do not, however, locate the event on the time axis, instead they modify the event notion expressed by the proposition. "Wait for three hours" is thus a quantified state or process consisting of the following phases: "start waiting", "keep waiting" and "stop waiting". Process propositions and unquantified activities behave in the same way as states. They overlap with or include their reference times: *When I opened the window, a cold wind was blowing - When I came home, Mary was playing the piano.* We should note, however, that processes and activities may not only overlap with reference points but also reference intervals: *From nine to ten I was watching television* may be interpreted as a situation where the watching of television extends beyond or includes the time span indicated by the time adverbial.

2.2 Punctual changes and reference time

Punctual changes, on the other hand, seem definitely to require reference points for their temporalization. They assert a punctual mutation phase, presuppose a specific anterior state, and imply a specific resultant state. We temporalize them via the punctual mutation phase, i.e., via a reference point: *At lo o'clock the light went on again.* We cannot temporalize the presupposed anterior state. The resultant state, however, can be quantified adverbially: *The light went on for a moment.* What we get in this case is a quantified event which must nevertheless be temporalized via the punctual mutation phase: *At lo o'clock the light went on for a moment.* I wonder whether punctual changes in English can be combined with time period adverbials, i.e. *from X to Y* adverbials with definite dates: *He was taken prisoner from the 1st of April till the 15th of May.*

2.3 Initially determined processes and reference time

Initially-determined processes differ from punctual changes in so far as the result phase can be isolated by the progressive: *He was looking away* denotes the averted gaze and (as a rule) not the avertion of the gaze. The central phase is the momentary mutation phase, usually conceived of as an indivisible unit or point. It is via this phase that such propositions are temporalized: *At that moment he looked away.*

2.4 Accomplishments and reference time

Accomplishments consist of a momentary initiation phase, a cumulative process phase and a momentary termination phase. These phases constitute an indivisible whole. We cannot isolate the individual phases. To say, for example, that "somebody had breakfast at 8 o'clock" clearly does not mean that only the initiation phase is being asserted. The speaker uses this phase to place the whole event on the time line. We can, of course, isolate the process phase with the progressive, but we are not concerned with this question at the moment. The normal case is that we relate the whole event (with all of its phases) to a reference interval with the same temporal extension as the event time. The appropriate time adverbials are "frame" adverbials: *He wrote a poem in ten minutes (from 10 to 12* etc.*)*.

2.5 Achievements and reference time

Achievements, finally, consist of a (cumulative) process phase, which is not directly asserted but implied or presupposed, and a punctual or momentary termination phase; it is via this phase that achievements can be related to reference points: *We reached the summit at 12 o'clock.* That the process phase is implied or presupposed becomes apparent by virtue of the fact that we can combine achievements with frame adverbials: *We reached the top of the mountain in six hours.*

3.0 Time adverbials

Before starting our discussion of the Past tense a few remarks on time adverbials may be useful.

3.1 R. Bäuerle's modification of H. Reichenbach's approach

We have so far adopted Hans Reichenbach's trichotomic approach towards the analysis of English tenses with speech time, reference time and event time as the fundamental three parameters on which to base our analysis.

Rainer Bäuerle[13] suggests a slight modification of this approach and also uses different terminology. He speaks of the evaluation time (Evaluationszeit) as the time where computation starts, in normal cases the speech time or a certain stage in the computation of a temporal expression such as, for example, a time adverbial. In *gestern vor einer Woche* "gestern" serves as an evaluation time from which we count back a week in order to arrive at what Bäuerle calls the "Betrachtzeit". A third parameter is the event time (Aktzeit).

3.2 R. Bäuerle's "Betrachtzeit" criticized by C. Fabricius-Hansen

The time adverbial usually indicates what Bäuerle calls "Betrachtzeit", which we cannot translate with Hans Reichenbach's "reference time" for the

---

13 Bäuerle (1979).

simple reason that, while in Reichenbach's approach event time and reference time coincide in several tenses, the event time in Bäuerle's approach is very often only a sub-interval of his Betrachtzeit.

The "Betrachtzeit", according to Bäuerle, is usually indicated by the time adverbial: e.g. *today, yesterday, last week* etc. The simple tenses (present, past and future) locate the "Betrachtzeit" on the time axis relative to speech time or identify part of the adverbially specified interval as the proper "Betrachtzeit" : *Hans kam heute.*[14]

This raises the question as to how the event time is located on the time axis in this model. Bäuerle says that every sentence contains, explicitly or implicitly, a frequency adverb and it is the task of these adverbs to locate the event on the time axis relative to the Betrachtzeit, i.e. to identify it as a sub-interval of the Betrchtzeit. Bäuerle would interpret the sentence

(ii) John cut his finger yesterday

as

(iii) John cut his finger (at least once) yesterday.

This paraphrase, however, leads to difficulties as pointed out by Fabricius-Hansen.[15] She maintains that sentence (ii) is to be looked upon as true if there is a (specific) sub-interval of 'yesterday' at which the proposition "John cut his finger" is realized. She characterizes event time as the truth interval of the proposition. This definite interpretation of event time allows or suggests the distinction between a Betrachtzeit, the time named as 'yesterday', and the event time or truth interval as a sub-interval of the Betrachtzeit. However, as Fabricius-Hansen points out, in an indefinite interpretation of the event time, i.e., by means of a frequency adverb (= *at least once*), the truth interval for the proposition "John cut his finger at least once" is not the event time but *yesterday*. These facts are not taken into account in Bäuerle's interpretation of event time and Betrachtzeit and Fabricius-Hansen's criticism of Bäuerle seems to be justified.

3.3 Various uses of frame adverbials (reference frames)

These considerations also shed some light on time adverbials. A threefold interpretation is apparently required for time adverbials such as *yesterday, last week, today, tomorrow* etc. They are, of course, orienting adverbials

---

14 ibid. (1979:53).
15 Fabricius-Hansen (1984:38-41).

in the sense that they identify an interval or time span on the time line. They undoubtedly name and mean this time span according to the lexical conventions of a specific language, English. But in different contexts they may be used to refer to several things:

(1) *Yesterday* for example can refer to "all time contained in yesterday":

It was foggy (windy, cold, etc.) yesterday may be paraphrased as "For all time contained in yesterday there was fog, wind etc."

We can further (2) use it to refer to "a specific time or sub-interval of yesterday": *Hans was absent yesterday* will usually be taken to mean "absent on a certain occasion."

Finally, we use it to refer to (3) "at some time (some sub-interval) of yesterday", which is the interpretation proposed for time adverbials of this type by Guenthner[16] and by Bäuerle.[17] The following sentence *(I'm sure that) somebody forced my door yesterday* apparently does not refer to a definite event time. What it seems to assert is that there must be some time contained in *yesterday* at which the proposition is true.

Here apparently the event time, because indefinite, cannot be looked upon as the truth interval for the proposition. In this case Bäuerle's interpretation does not seem to be correct.

What is important in connexion with the interpretation of time adverbials such as *yesterday* is the possibility of taking them to refer to one of the specific, though textually unidentified, sub-intervals they can be thought to consist of. Such an interpretation seems also to have been envisaged by Fabricius-Hansen.[18] And I feel that this is correct since, as M. Bennett and B. Partee note, we can interpret sentences in the Past tense without time adverbials as referring to a specific contextually supplied time:

I forgot to turn the stove off.

4.0 The function of the Past tense

In this section we are concerned with the interpretation of the Past tense. We adopt Reichenbach's approach and retain his terminology as far as possible. We distinguish speech time, the primary orientation axis. We distinguish

16 Guenthner (1979).

17 Cf. Bäuerle (1979:45):" 'Definit' sind also Sprechzeit und Betrachtzeit, die Aktzeit bleibt innerhalb des von der Betrachtzeit gesetzten Rahmens indefinit."

18 Fabricius-Hansen (1984:63): ... "es ist damit zu rechnen. daß mit "rahmenbildenden" oder "randdefiniten" Betrachtzeit-Adverbialen (...) wie *heute,gestern, dieses Jahr, letzten Sommer* in etwas laxer Weise auf kontextuell bestimmte Teilintervalle des jeweils bezeichneten Intervalls referiert werden kann."

between primary and secondary speech time in discussing direct, indirect and substitutionary or indirect free speech. We further adopt Reichenbach's reference time as the time in relation to which an event time is located on the time axis. The event time, finally, is the time at which a proposition is realized. It may be necessary to distinguish a fourth parameter, the adverbially specified time, which may or may not be identical with the event time or reference time.

4.1 The Past tense in isolated sentences interpreted as a search instruction

We use the Past tense in textually isolated sentences without time adverbials. Let us return to the example already mentioned, *I forgot to turn the stove off*, adapted from B. Partee (1973). To interpret this sentence in the way proposed by A.N. Prior or Rescher/Urquart and other tense logicians[19] would be to miss its communicative import. This sentence clearly refers to a specific past time, which is supplied by the speech situation or "Redehintergrund": We call this time 'reference time' and say that in the present case it is identical with the event time.

We do not wish to suggest that it is the Past tense that supplies this definite time and we prefer to say that the Past tense introduces the whole span of time anterior to speech time as a reference frame from which the sentence-in-context chooses a specific sub-interval as reference time. More specifically still, we propose here to describe the function of the Past tense as a search instruction which may be provisionally paraphrased as follows: "Look out for a co-textually or contextually given specific time in the past, i.e. anterior to and not including speech time (in some cases reference time) and relate the untensed or temporally unspecified proposition to it as simultaneous."

This instruction would apparently exclude any indefinite interpretation of simple Past tense sentences and the question is whether *somebody stole my car* could be used and understood as meaning that the theft occurred at some time in the past. We must leave this question open for the moment.

We further maintain that a simple Past tense sentence would remain uninterpretable if a specific reference time were not supplied either co-textually or contextually.[20]

---

19 A useful discussion of various tense-logical approaches can be found in Rainer Bäuerle(1979:3ff.).

20 Cf. Fabricius-Hansen (1984:12f.).

4.2 The Past tense as a search instruction in combination with adverbials

The Past tense is very often accompanied by time adverbials. Our interpretation of the Past as a search instruction would work perfectly well in such cases:

John lost his watch yesterday

would have to be represented as

Yesterday (Past (John lose his watch)).

The interpretation of the sentence by the hearer could be described as follows. He would register the event type of the tenseless proposition: "John lose his watch" as a punctual change. The Past tense would instruct him to look for a suitable reference time in the past (a) sentence-internally or, (b) co-textually or contextually supplied. The sentence-internal time adverbial is *yesterday*, a time span which cannot be filled by the momentary or punctual mutation phase referred to in our sentence. Consequently, a sub-interval of *yesterday* must be envisaged, which raises the question of whether a specific or an indefinite sub-interval is to be choosen. The information to be conveyed is apparently that the loss occurred at some time yesterday and not at a definite time. This also describes the conditions under which this sentence would be considered to be true. An indefinite interpretation seems the most appropriate solution.

Unquantified state and process propositions require an interior reference time. To say *At 8 o'clock it was raining hard* means to say that the reference time 8 o'clock is surrounded by the event "raining". This event is not bounded, yet we know and assume on the basis of our knowledge of the world that this process is of limited duration, that is, is temporary. We further know that the event "rain" is also predicated to extend into the future after 8 o'clock.

We represent this sentence as

At 8 o'clock (Past (it rain fast)).

To process this sentence the following steps are necessary: We start with the innermost bracket and ascertain its event type as a temporary or variable unquantified process. We know it to be a rule of English that events of this type require an interior reference time or point. The Past tense instructs us to look for a reference point or time under the conditions we formulated for this tense. The only reference point availible is provided by the sentence-interior time adverbial, which, if complemented by the larger context, complies with our search conditions. The Past tense relates the proposition to this reference point and our interpretation ends by placing the reference point inside the event. This is the only correct interpretation of the above sentence.

Accomplishment predications or accomplishment propositions can be temporalized in different ways. We know that they consist of three distinct phases (initiation, cumulative process, termination), which are predicated as a whole: *From ten to eleven I wrote a letter to my father*. This can be represented as

From ten to eleven (Past (I write a letter to...)).

Again we start with the innermost bracket and assess the event type it represents. The proposition is undoubtedly an accomplishment. In order to be placed on the time line it requires a time interval. The Past tense looks out for a reference time in the past. The time adverbial supplies an interval but lacks any indication as to its position on the time line. The additional information we need would have to be supplied by the speech situation if we want to make the sentence communicatively effective.

We have already pointed out that accomplishment propositions can, on the surface, be combined with point-of-time adverbials: *At 11 o'clock I wrote a letter*.[21] The reference time in this case would not simply be 11 o'clock, but an interval starting at this point.

The examples discussed so far suffice to demonstrate that the first step in our analysis, the identification of the event type the proposition in the innermost bracket belongs to, is an indispensable operation because it determines the type of reference time (point, closed or open interval, etc.) to be looked for.

We can sum up the preceding discussion as follows: The ultimate goal to be attained by processing a fully tensed sentence, i.e. a sentence containing tense and time adverbials, is to locate the event reported by it in time. It is the determination of event time we try to achieve. We may distinguish several steps. We demonstrate them by analyzing *John lost his watch yesterday*, which we render provisionally as <u>Yesterday (Past (John lose his watch))</u>.

---

21 Cf. Fabricius-Hansen (1984:10).

We consider the time of speech (S) as given by the pragmatic context (the act of communication).

- Step 1 will determine the event notion or situation type signalled by the untensed sentence concept. "John lose his watch" is clearly a punctual change requiring a punctual reference time to be located in time. This reference time contains or includes the event or is identical with it: $E \subseteq R$.

- Step 2 consists of analyzing the information supplied by tense. The Past tense instructs us to look for the reference time in the time sphere anterior to speech time ($R<S$) but to start our search sentence internally, which leads us to

- Step 3, the computation of the time adverbial in relation to the information gleaned from tense. *Yesterday* specifies a time interval anterior to S and is therefore compatible with the search instruction of the Past tense. The time adverbial specifies the information supplied by tense. But there remains the question of how the time adverbial tallies with the punctual reference time. The inevitable conclusion is that the punctual reference time must be included in the time interval referred to by *yesterday*: ($R \subset TA$). This leads on to

- Step 4, the question of whether the reference time contained in the reference frame "yesterday" should be looked upon as a definite or indefinite time. The lexical content of the sentence concept ("lose one's watch") suggests that an indefinite reference time is asserted: $\exists t(t = R \wedge t \subset TA)$. The result of our analysis is that the event (John lose his watch) occurred at an indefinite (and unspecified) time contained in the reference interval "yesterday".

We can sum up our analysis in the following formula:

$\exists t\ (t = R \wedge t < S \wedge t \subset I \langle yesterday \rangle \wedge t \supset E \langle John\ lose\ his\ watch \rangle)$.

The formula seems to supply all the information necessary for the reconstruction of the original text. We may even simplify it for this purpose, "yesterday" of course left unanalyzed:

$\exists R\ (R \subset yesterday \wedge R < S \wedge R \supset E \langle John\ lose\ his\ watch \rangle)$.

The reconstruction of the text would start with the information $R \supset E$, which suggests the simple form, the information $R < S$ enforces the Past tense with which the time adverbial harmonizes. *John lost his watch* can be retrieved from the formula.

We try to transpose another formula into a natural language version:

(R = 10 p.m. yesterday $\wedge$ R < S $\wedge$ R $\subset$ E<John sleep>).

The information R $\subset$ E suggests the progressive, R < S indicates the Past tense, with which the TA harmonizes. The formula translates into *Yesterday at 10 p.m., John was sleeping.*

We shall briefly discuss the suggestion made by Bäuerle[22] that Reichenbach's reference point can be dispensed with in the analysis of the simple tenses, present and past. Bäuerle is, of course, right in saying that in these tenses reference point and event time may fall together. This is so because Reichenbach was not aware of, and therefore did not take account of, the level of event type. However, to temporalize a state would clearly be impossible on the basis of the two parameters of speech time and event time only, for the simple reason that in the case of states the event time is not defined. To relate a state to the time line requires a time point interior to the event, i.e. inside the state. We also clearly need a reference time for the analysis of the simple tenses, and especially, of the difference between Present Perfect and Past, as Bäuerle recognises only too well.[23] A further argument in favour of Reichenbach's trichotomic approach is the secondary or relative tenses.

4.3 The Past tense in complex sentences (containing temporal clauses)

The Past tense is also used in complex sentences. Since Temporal clauses are discussed in detail in the second volume of this collection of essays, a few remarks must suffice here to illustrate our point. Temporal clauses function as time adverbials and may supply (a temporal frame for) the reference point for the main clause. However, in order to be able to do so, the events temporal clauses refer to must be located on the time line in a definite way. If this can only be done via a reference point as in the case of states and processes, the inevitable conclusion is that the temporal clauses presuppose their assertion as independent clauses. A sequence of sentences such as the following:

John was looking out for Mary. When she turned up, he waved his hat.

will probably have to be interpreted as the sequence

---

22 Bäuerle (1979:49): "...für die einfachen Tempora scheint mir der "point of reference" schlicht redundant zu sein.

23 ibid. (1979:52).

(1) John was looking out for Mary.
(2) Mary turned up.
(3) When she turned up, he waved his hat.

If we accept such an analysis, we could say that the temporal clause takes a given event time supplied by the preceding sentence in order to establish a reference time for the main clause.

One could, however, perhaps also say that our analysis contains a considerable amount of redundancy. We, therefore, propose to analyze the temporal clause via a secondary reference time (r) which will serve or function as the principal reference time (R) for the main clause. If we want at the same time to take account of the back- and foregrounding function of the clauses[24] we can indicate the background function of the temporal clause graphically by placing it above the time line:

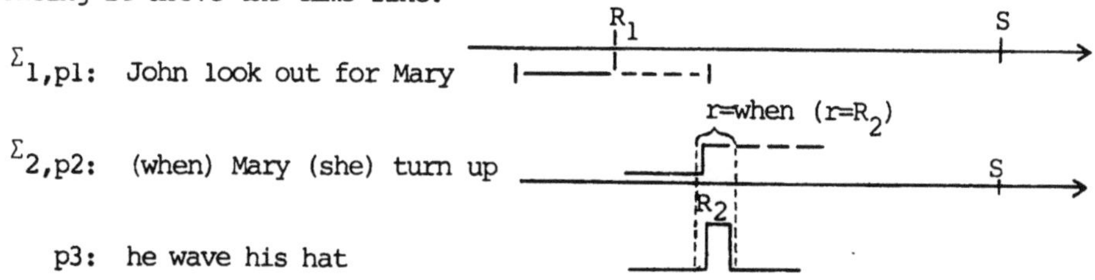

We indicate in the diagram that p2 is backgrounded. That temporal clauses need not necessarily be backgrounded in time-comparing sentences or in narrative sequences is convincingly demonstrated in Couper-Kuhlen (in volume 2).

Particularly interesting are temporal clauses which are non-factive: *He tore the letter before he (had) read it.* These will be discussed extensively in Hamann (volume 2).

4.4 The Past tense in sentence sequences: the progression of the reference time

The Past tense also occurs in sentence sequences, i.e. in stories. For this use the term "narrative Past" has been coined and various theories have been proposed to explain this function. Käthe Hamburger propounds the theory that the narrative Past is not a tense at all but is primarily connected with the fictional character of the stories told. We shall not discuss this theory

24 This topic was introduced into our discussion by Elizabeth Couper-Kuhlen.

here nor the criticism it evokes. The question we are interested in is how stories develop in time and which signals control the analysis of a text in its temporal structure. Particularly interesting is the question as to why certain sentence sequences are an iconic representation of a sequential series of events, why other sequences of sentences describe simultaneous events, i.e. situations, and why other sequences report temporally unordered events. We might classify these functions as narration, description and reporting. The latter term would, of course, have to be understood in a highly technical sense, viz. as just naming the events or states of affairs irrespective of their temporal order. We therefore distinguish three modes of discourse: the narrative, the descriptive and the reporting mode.

We propose to look upon these modes as being brought about by linguistic signals. This is not a new approach. The narrative mode was thought to be brought about by so-called narrative clauses. A narrative clause was understood to move the story on in time by one step and the narrative character of such clauses was thought to reside in the event type or event notion realized by the proposition.

4.4.1 W. Zydatiß and the time relation between successive sentences in sentence sequences

To my knowledge, the first to propound such ideas at some length was Wolfgang Zydatiß.[25] He distinguishes four possible relations between successive sentences: (1) succession, which in order to be brought about, requires "bounded" propositions (*When he saw me, he laughed loud*, - (2) simultaneity, which requires "unbounded" propositions (*While Moira was playing the cello, I watched television*), - (3) regress, flashback or retrospect, which can in most cases be signalled by tense and aspect choice in the clauses involved (*John had produced a strip of paper and was studying it with an ink-pencil between his fingers*), - and (4) co-incidence (for the German term "Inzidenz"[26]; please note that E. Koschmieder used the term "Koinzidenz" to denote the performative use of speech-act verbs[27]): *Mary was playing the piano when John knocked on the door.*

---

25 Zydatiß (1976:233).
26 ibid. (1976:233).
27 Koschmieder (1945:22-29).

As these examples show, these four relations can be reduced to two: viz. succession (Zydatiß' cases 1 and 3) and simultaneity (or overlapping) (Zydatiß' cases 2 and 4). As mentioned above, a third case must be distinguished, the case where two or more successive sentences refer to temporally unordered events.

Zydatiß is realistic enough to notice that these different relations between successive sentences depend not only on the two types of bounded and unbounded situations, but also on what he calls "Ordnungs- und Kausalbeziehungen"[28] between the events.

4.4.2 E. Hinrichs and the progression of reference time

Another interesting attempt to deduce the progressive organization of events in successive sentences from the "Aktionsart" of the propositions involved has been made by Erhard Hinrichs.[29] In discussing the following sentence - *He took his clothes off and took a shower* - he first considers the idea that the successive ordering of the two events is somehow to be attributed to the Past tense in the second clause. This "anaphoric" function of the Past tense would have to be described as the indication of posteriority of the event reported in the following sentence in relation to that of the preceding one.[30] This means attributing to the Past tense two diverging functions: in isolated sentences it would signal anteriority to the time of speech, in successive sentences it would signal posteriority to the preceding reference or event time. Hinrichs drops this idea later, pointing out that it is the "Aktionsarten" of the verbs, ultimately of the clauses involved, that are responsible for the temporal relation between events in successive sentences:

> "Wie zwei Ereignisse zueinanderstehen, dies hängt wesentlich von den 'Aktionsarten der Verben' ab, mit denen die Ereignisse bzw. Sachverhalte beschrieben werden." 31

Hinrichs bases his theory first of all on the four event types proposed by Vendler: activities, states, achievements and accomplishments, which in his opinion are responsible for the temporal ordering of events in successive sentences, secondly, on Hans Reichenbach's trichotomic analysis of the English tenses and, thirdly, on Kamp's proposals for discourse representation

28 Zydatiß (1976:233).
29 Hinrichs(1981).
30 ibid. (1981:56).
31 ibid. (1981:58).

structures. Hinrichs formulates the following rules or presupposes the following regularities:

1) An event (e_1) expressed in a sentence (s_1) with the simple Past of an achievement or accomplishment proposition, follows the reference point r-1 (in the discourse structure D-1) and introduces a new reference point (rl + 1) posterior to (e_1).[32]
2) "Activities" and "states" overlap the reference point already reached in the analysis. They do not introduce new reference points, which means that states or activities in successive sentences are related to one and the same reference point. They are simultaneous.[33]
3) Accomplishments and achievements are contained in the reference point; states, activities and progressives overlap or include the reference point.[34]

We see that the temporal relation between the events of a story, according to Hinrichs, is signalled by two classes of propositions, accomplishments and achievements on the one hand, and states and activities on the other hand. There is little difference in this respect between Zydatiß and Hinrichs. And both are aware of the fact that it is not "Aktionsarten" or "event types" alone that are responsible for the progression of the reference point. As already pointed out, Zydatiß recognizes that causal and pragmatic relations between events in extra-linguistic reality play a role in this respect and Hinrichs points out that it is the choice of the subject in two successive sentences and of locative adverbials that contributes to the specification of the temporal relation between the events referred to.[35] This is a valuable suggestion because it is a first step towards taking account of several sentences in determining their temporal order. It also seems correct to say that accomplishments and achievements are contained within, i.e. are simultaneous to or co-extensive with, the reference point, whereas states and activities overlap or include it. But Vendler's list of event types is not exhaustive and the question remains as to which type of reference time these event types require in order to be placed on the time line. This question, i.e. the tensing potential of the event types, does not seem to have been satisfactorily dealt with yet.

---

32 Hinrichs (1981:61).
33 ibid. (1981:64).
34 ibid. (1981:78).
35 Zydatiß (1976:233 ff.) and Hinrichs (1981:80).

4.4.3 B. Partee and the automatic progression of reference time

B. Partee's approach is based on ideas outlined in Hinrichs (1981), which she adopts and extends.[36] Partee also uses Kamp's discourse representation structures and Vendler's four event types. She endorses the view propounded by Zydatiß and Hinrichs that certain event types, i.e. Bach's instantaneous and extended events,[37] move the action forward in time, whereas states and processes do not, but instead describe how things are at the time of the last mentioned event.[38] She maintains that *when*-clauses in the Past tense trigger the up-dating of the reference time to a new value and adopts Hinrichs' view that event sentences are contained in their reference times while states and processes overlap the reference times. Reference times, incidentally, are looked upon as "protracted events".[39]

Against this background of general assumptions she analyzes the following sequence (Example 16 in her paper) -

1. Mary turned the corner.
2. When John saw her,
3. she crossed the street.

by the following diagram abstracted from a more detailed discourse representation structure.[40]

$r_o < r_1 < r_2 < r_3 <$ <u>now</u>

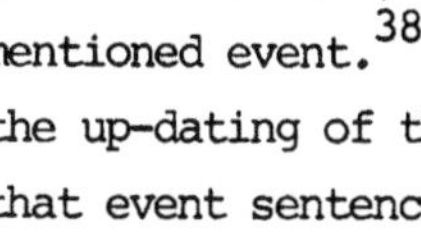

The symbol "<" is to be read as "completely precedes", r_o is the reference time we need for the beginning of a story. In hearing sentence 1 the listener is apparently expected to assume a reference point in the (fictional) past and to relate the event referred to in sentence 1 as simultaneous to this reference time r_o, as indicated by the circle round e_1. Sentence 1 is an event sentence and therefore introduces a new reference point, r_1, just after r_o. The following sentence starts with a *when*-clause, which, according to B. Partee, updates the reference point to r_2. The temporal sentence as a whole is

36 Partee (1984:243ff.).
37 Bach (1981:63ff.).
38 Partee (1984:253).
39 ibid. (1984:258).
40 ibid. (1984:260).

apparently thought to relate both events referred to in its clauses, e_2 and e_3, to this reference point. Note that e_2 and e_3 are surrounded by a circle. The main clause of the sentence introduces a new reference point, r_3, which is earlier than speech time and brings the story to its end.

Unclear in this analysis is why the introduction of r_1 should be necessary, since it seems to have no function whatever. In the preliminaries to the analysis it is expressly stated that events (accomplishments and achievements) are contained in their reference points. If, on the other hand, a new reference point is introduced by the following *when*-clause and the event of this clause is contained in the newly introduced reference point, it clearly follows that e_1 fully precedes e_2 and no r_1 is necessary to seperate them. There is no information whatever that could be gleaned from r_1.

Barbara Partee adopts another of Hinrichs' views, namely the principle that states and activities need not necessarily be understood to overlap the preceding event (and its reference point) but should be related to a new reference point introduced by the preceding event. Hinrichs apparently believes that this principle can be substantiated by discourses such as the following,[41] reprinted in B. Partee:[42]

> Jameson entered the room (e_1), shut the door carefully (e_2), and switched off the light (e_3).
> It was pitch dark around him (s_1), because the Venetian blinds were closed (s_2).

Hinrichs and Partee point out that states necessarily contain or include their reference times. The last event contained in a reference time is "switch off the light" (e_3). The following state "be pitch dark around him", however, certainly does not overlap with e_3. It is for this reason that Hinrichs postulates the principle of the introduction of an additional reference point, automatically introduced by the last event sentence (e_3). Hinrichs' analysis, it is true, is fundamentally correct: the state s_1 must be conceived of as overlapping with a new reference point later than e_3. But should this new reference point be supplied by an automatic progressive principle which, as we have seen, seems in some cases to lead to a good deal of redundancy?

---

41 Hinrichs (1981:66).
42 Partee (1984:254).

4.4.4 Our proposal compared with B. Partee's

Perhaps a more realistic analysis can be found. I think that it is undoubtedly correct to look upon a story as organized as a series of events related to temporally ordered reference times. A narrative text contains several modes of presentation: narration of successive events, description of situations and the explication of macro-events (a party, an accident etc.) through the unordered enumeration of sub-events. As indicated above, we believe that these modes are somehow signalled by the text itself and its background.

As far as the progression of events and reference points is concerned, there is of course a general iconicity of the utterances following one another relative to the events following one another. A sentence following another may therefore tend to be interpreted as referring to one event following another: *He had a shower, undressed, and went to bed* will probably lead to the question of whether the person referred tc had a shower with his clothes on. But in general, texts supply preciser information than that to be gleaned from this general iconicity of utterance sequences.

But not only should we try to avoid the redundancy found in Hinrichs and Partee, we should also try to find a unified and general analysis for the Past tense. We indicated that Hinrichs' analysis leads to at least two fundamentally contrastive functions: in isolated sentences the Past tense signals simultaneity to a reference time anterior to speech time, in sentence sequences the Past tense signals simultaneity with a reference time posterior to a contextually or co-textually given time. In isolated sentences the principal "vectorial" function of the Past tense is anchored to speech time, in sentence sequences it is anchored to a preceding reference time.

In my opinion, a unified function of the Past tense can be maintained, if we understand this tense as a search instruction adapted to several different cases.

I propose the following general instructions:

(1) Look for a definite reference time in the past (anterior to speech time) signalled
 (a) sentence-internally (e.g. by a time adverbial), or
 (b) by the context (speech situation, co-text etc.)
 and put the event or situation referred to in the sentence to it in a relationship of simultaneity, i.e., as contained in it or including it.

(2) In a sequence of sentences, put the event of the following sentence in a relationship of simultaneity to the reference time established by the preceding sentence.
If temporal incompatibility, causal or pragmatic relations between the two sentences prevent this, assume a new reference time posterior to the last one and have it include the event referred to, or let it be included in it.

The search instructions ascribe three values to the past tense: (1) anteriority to speech (or reference)[43] time - (2) simultaneity of the event referred to with the reference time (inclusion or overlap) - (3) the definite pastness of the reference time.

These search instructions seem to interpret the temporal structure of narrative texts correctly.

Let us test it on some of the texts discussed in Hinrichs and B. Partee. First we shall look at the text of Example (16) in B. Partee:[44]

(16) (A) Mary turned the corner.
(B) When John saw her,
(C) she crossed the street.

The Past tense in (A) leads to the assumption of a current reference time (r_o) in the (fictional) past with which to start the story. The following *when*-clause, (B), seems at first to suggest that the event "John see her" is simultaneous to r_o. But the linguistic context, particularly the simple form in (A), which conceives of "Mary turn the corner" as a whole or blob, is such that simultaneity of the event in (A) and (B) appears unlikely, if not impossible. The decoding intelligence assumes a new reference time (r_1) posterior to r_o and relates the event in (B) as simultaneous to it. The next step in the decoding process is to look for a reference time for the Past tense in (C), i.e. for "she cross the street", which, on the basis of our search conditions, will be found in r_1 to which the event in (C) is related as simultaneous.

There seems to be some difficulty lurking in the *when*-clause. The event in (C) is not conceived of as strictly simultaneous with the preceding event, or the preceding reference time, but as contained in it together with the event in (B) and immediately successive to it. The temporal analysis of *when*-clauses seems to depend on a number of factors, particularly on the event types of the

43 Anteriority to reference time may occur in complex sentences.
44 Partee (1984:258).

propositions involved, as Hinrichs[45] recognizes.

Another interesting text discussed by B. Partee is the following, Example (14) in her paper[46]:

> (A) John got up (e_1), (B) went to the window (e_2), and (C) raised the blind (e_3). (D) It was light out (s_1). (E) He pulled the blind down (e_4) and (F) went back to bed (e_5). (G) He wasn't ready to face the day (s_2). (H) He was too depressed (s_3).

Our interpretation can, as indicated, be looked upon as some sort of decoding process: Sentence (A) leads to the assumption of a temporal starting point (r_o). Sentence (B) tries to relate e_2 as simultaneous to r_o. This is prevented by the relation between e_1 and e_2. They are clearly temporally incompatible. On the basis of our assumption of the unidirectional sequence of events, the listener assumes a new reference time (r_1) posterior to r_o. This reference time includes e_2 . The decoding process moves on to sentence (C). The Past tense tries to relate e_3 to r_1. Temporal incompatibility between e_2 and e_3 again prevents this. The listener assumes a new reference point (r_2), posterior to r_1, and relates e_3 as simultaneous to it. In processing sentence (D) we can interpret it as either belonging to the deictic field of the narrator or to that of the narrated person, i.e., as substitutionary speech. We opt here for the former interpretation: the Past tense in (D) relates the state referred to in this sentence ("it be light out") to the reference point already established ($=r_2$) as surrounding it. Partee would here have a new reference point (r_3) introduced by sentence (C) posterior to r_2, and the state s_1 would overlap it (r_3). However, this analysis does not explain how we can guarantee that s_1 also overlaps with r_2 as, apparently, our understanding of the text requires. In proceeding to the decoding of sentence (E), we see that we cannot relate its event to the last established reference time. The event of sentence (E) is incompatible with e_3; a new reference time posterior to r_2 has to be set, (r_3), and e_4 related to it as being included in it. The next step in the decoding of the text again first tries to relate e_5 in (F) to the last established reference time as simultaneous with it. Temporal incompatibility prevents this. A new reference time posterior to r_3 has to be established, (r_4), and e_5 included in it. Sentences (G) and (H) are comments on the situation by the narrator and relate states s_2 and s_3 to r_4 as overlapping it.

---

45 Hinrichs (1981:82ff.).
46 Partee (1984:253).

Simple as this analysis may be, intuitively at least, it appears to be correct. But the text also permits a different interpretation. Sentences (D), (G) and (H) could be interpreted as "substitutionary speech" or "indirect free speech". This would necessitate the location of their event and reference times in a secondary deictic field. We return to this question later.

Particularly interesting is the text in Example (15) in B. Partee's paper.[47] As pointed out above, Hinrichs uses this text to substantiate his claim that an event sentence should be understood to automatically introduce a new reference point posterior to the one it is simultaneous to.[48]

In the following it is argued that this automatic introduction of a new reference time by "event" sentences is not a necessary prerequisite for an account of the temporal relations in the text. The text runs as follows:

> (A) Jameson entered the room (e_1), (B) shut the door carefully (e_2), and (C) switched off the light (e_3). (D) It was pitch dark around him (s_1), (E) because the Venetian blinds were closed (s_2).

It is obvious that s_1 in sentence (D) does not overlap the preceding event. We apparently need a reference time posterior to that of sentence (C) to allow s_1 to overlap it, i.e., to place the listener inside s_1. This should not be done by a general rule which introduces a new reference point after every "event" sentence (whatever this may be). The introduction of a new reference time can in this case also be understood as the result of a calculation on the basis of our search instructions. The decoding apparatus (hearer) recognizes that state s_1 is the result phase of e_3. If we stick to our rule that unquantified states must overlap with their reference points, it follows that a new reference point (r_3) must be introduced. State s_2 in sentence (E) is also understood to overlap r_3, but on the basis of his "world knowledge" the listener has no difficulty in construing the situation in such a way that s_2 overlaps all the reference points and events referred to thus far.

The shifting of the reference time need neither be attributed to the Past tense nor looked upon as automatically caused by "event" sentences, but would in each case be the result of a calculation process comparing every new sentence with its decoded context and making use of all other pertinent background information available to the decoder.

47 Partee (1984:254).
48 Hinrichs (1981:66f.).

From this point of view it seems rather unlikely that this decoding process could ever be completely formalized or mechanized. The information accessible and made use of for the decoding of narrative texts does not seem to be storable and retrievable in a formalized and mechanized way.

As pointed out above (p.193) E. Hinrichs entertained the idea for a while that the anaphoric function of the Past tense could be described as signalling posteriority to a reference or event time supplied by the preceding context: "Aus den Sätzen (79) - (82) könnte man den Schluß ziehen, daß die anaphorische Funktion des Past tense darin besteht, jeweils Nachzeitigkeit zum im Vordersatz genannten Ereignis auszudrücken."[49] Hinrichs seems to abandon this idea later and substitutes for it his principle of the automatic shifting of the reference point by "event" sentences (accomplishment and achievement propositions). The idea of a "posteriority" function of Past tense sentences in sequences, however, seems to have been taken up by Fabricius-Hansen.

4.4.5 Event sequences in C. Fabricius-Hansen's approach

She discusses the following sequence, Example (52) in her manuscript[50]:

(d) Er fing Botha die Arbeiter weg,
(e) bedrohte sie mit Folter und Kerker, um Bothas Geheimnis auf die Spur zu kommen ....

Fabricius-Hansen attempts to interpret tensed sentences against what she calls "Bewertungskontexte" (evaluation contexts) made up of three indices: the speech time index (t_o), the "Betrachtzeit"index (T_b), which can in some cases be equated with our reference time, and the "co-time" index (T_{ko}), i.e. any time expression supplied by co-text and context.

Tensed sentences can be anchored to speech time or, if they are part of a sentence sequence, to co-times. In the latter case, a sentence without a time adverbial will make use of the time adverbial of the preceding sentence. It will be taken over as "Betrachtzeit" (t_b), i.e. as the second index of the evaluation context. If there is no time adverbial functioning as (t_b) in the preceding sentence, the event time or truth interval of the preceding sentence takes over this function so that the sentence under analysis will be evaluated on a context such as (t_o, T_k, T_{ko}) where T_k is the event time or truth interval

49 Hinrichs (1981:56).
50 Fabricius-Hansen (1984:91).

of the preceding sentence. The untensed proposition underlying the sentence under analysis, however, will take over T_k as its evaluation time. The evaluation context of the untensed proposition will, therefore, be (t_k, T_k, T_{ko}). The truth-conditions C. Fabricius-Hansen proposes for an untensed proposition (ϕ) in such contexts say that (ϕ) is true if a truth interval of (ϕ) either surrounds (t_k), i.e. is simultaneous with (t_k), or immediately follows it.[51]

In applying this to her text, Fabricius-Hansen says that as sentence (d) has no time adverbial, it is its event time (t_d) or truth interval that is taken over as "Betrachtzeit" (t_b), i.e. as second index by the following sentence (e). The untensed proposition underlying this sentence, however, would also take (t_d) over as its first index, that is as evaluation time, supplying it with the evaluation context (t_d, T_d, T_{ko}). The truth conditions for propositions in such contexts allow the event time of (e) immediately to follow (t_d). And as the event time of (e) will in its turn serve as "Betrachtzeit" (T_b) for the next finite sentence and as evaluation time (first index) for its untensed proposition, a sequence of events can be described by successive Past tenses. But since, as pointed out above, the truth conditions allow succession as well as simultaneity for sentences in sequence, the question arises as to which factors control the choice between these two alternatives. According to Fabricius-Hansen it is the event notions, in her terminology the "Aktionsart" of the propositions under consideration, that controls this choice. I have pointed out above that this cannot be the whole truth.

5.0 The Past tense in the various modes of speech and thought representation

The Past tense also occurs in the various modes of speech and thought representation. In most of these cases the Past tense is a transposed Present tense. We shall discuss here only the case of substitutionary speech (sometimes called free indirect speech) and substitutionary pereception.

Examples containing substitutionary speech are the following:

(a) Tomorrow was his birthday.
(b) He was going to Berlin tomorrow.
(c) He pulled the blind down and went back to bed.
He wasn't ready to face the day. He was too depressed.

---

51 Fabricius-Hansen (1984:93): "(I) Sei ϕ eine zeitlich neutrale Proposition. (a) ϕ ist wahr (falsch) an einem Zeitkontext vom Typ (t_i, T_i, T_k), wenn ein Wahrheitsintervall von ϕ t_i umgibt oder innerhalb eines relativ kleinen inhomogen nach t_i liegenden Intervalls positioniert ist, je nach den Eigenschaften (u.a. der Aktionsart) von ϕ."

Examples of substitutionary perception are the following:

(d) John got up, went to the window, and raised the blind. It was light out.

(e) He left the room and went to a window on the first floor and looked out. The street was quiet and no cars were running. The tracks were buried under snow. 52

5.1 Various interpretations

Various interpretations and approaches have been proposed.

5.1.1 B. Fehr

The first to deal extensively with these phenomena was Bernhard Fehr. It was he who suggested the terms 'substitutionary speech' and 'substitutionary perception'. He characterizes substitutionary speech as follows: "This is the meaning of 'Erlebte Rede' - the reporter experiencing the actor's speech, speaking in the actor's name, substituting his own voice for the actor's although still aware that he is the first person and the actor the third. So it is partly the actor speaking through the reporter and partly the reporter speaking through the actor. It is just this mellowing twilight enveloping an utterance that constitutes one of the charms of substitutionary speech."[53]

He further points out that substitutionary speech is part of a more comprehensive stylistic device, that of substitutionary narration and description to which a second technique belongs, substitutionary perception, characterized as the actor lending his sight or audition to the reporter and the reporter running in on the actor's vision or audition. Such passages are usually introduced by what B. Fehr calls a "perception indicator" pointing to or 'opening' a new perception area (e.g. "everybody listened attentively", "he looked out of the window", "he raised the blind" etc.). Substitutionary perception is usually rendered in the progressive and for very good reasons. It must not be mistaken, however, for a verbal rendering of the thoughts of the actors. According to Bernhard Fehr, substitutionary perception produces "the illusion of receiving a direct verbal replica of visions and auditions not

52 Wright (1972:281).
53 Fehr (1944:265).

yet affected by the stream of reflections"[54] and should therefore be distinguished from "parole interieure".[55]

5.1.2 G. Rauh

Gisa Rauh interpreted substitutionary speech and perception aptly as the interpenetration or amalgamation of two deictic fields, of the orientation areas of the narrator and of the narrated person, and made an attempt to represent this on the basis of an extended hypersentence analysis.[56] However, the graphical devices this analysis yields are too unwieldy to be used for practical language teaching or literary propaedeutics, the spheres we are primarily interested in.

5.1.3 G. N. Leech and M. H. Short

We should in this connection perhaps also mention the attempt made by G. N. Leech and M. H. Short to arrange the various modes of reporting speech and thought as a cline of narrator's control of the report starting with the narrative report of action or speech act (apparently in total control of the narrator) and sliding down via indirect speech, free indirect speech (apparently Fehr's substitutionary speech) and direct speech (report only under partial control of the narrator) to free direct speech (narrator not in control of the report at all).[57]

It is to be regretted that B. Fehr's substitutionary perception is not taken account of in an otherwise useful survey.

5.1.4 G. Steinberg

The most extensive study of substitutionary speech (Erlebte Rede) comes from Günter Steinberg.[58] Especially useful are his attempts to formulate rules for the transposition of the tenses, the personal pronouns and the use of adverbials and names. Any attempt to describe substitutionary speech and perception will have to take account of the information supplied by Steinberg and to make allowance for the fact that the various modes of speech and

---

54 ibid. (1944:271).
55 ibid. (1944:272).
56 Rauh (1978, esp. Chapter IV).
57 Leech/Short (1982, chapter 10, esp. p.324).
58 Steinberg (1971).

thought and perceptual representation involve quite a number of subvarieties.[59]

5.2 Substitutionary speech on the basis of truth conditional semantics

Finally, mention must be made of the attempt to study substitutionary speech (Erlebte Rede) on the basis of a truth conditional semantics. This has been undertaken by Catherine Fabricius-Hansen.[60]

She starts by proposing that sentences containing a tense and a time adverbial be analyzed in two different ways, viz., to consider either the tense or the time adverbial as the highest temporal expression which then has the other temporal expressions within its scope. This would lead in the case of the following sentence (Example (20) in her manuscript)

(20) Anne took her exam last year

to the following two interpretations:

(20) (a) Past (last year (Anne take her exam))

(20) (b) Last year (Past (Anne take her exam)).

In (a) the Past tense supplies as a first possible reference time the whole sphere of the past exclusive of and relative to the speech time. The time adverbial specifies a sub-interval (last year) as a reference frame in which to look for an event time for the remaining proposition (Anne take her exam).

In the (b) version the result of processing the sentence is the same: The time adverbial specifies a definite past interval as reference time and the Past tense with its general contribution turns out to be redundant: *last year* is the reference frame in which the proposition (Anne take her exam) has to be located.

However, not all sentences containing a past tense and a time adverbial allow these two interpretations. Narrative texts in English do contain sentences such as

(21) Tomorrow his mother came to visit him.

If we look at the two interpretations proposed for such sentences

(21) (a) Tomorrow (Past (his mother come to visit him))

(21) (b) Past (tomorrow (his mother come to visit him))

---

59 Cf. Steinberg (1971:165-376), cf. also Leech/Short (1982:322-351).
60 Fabricius-Hansen (1984:96ff.).

we see that in (21a) the time adverbial (*tomorrow*), as the highest temporal predicate, would be anchored to speech time and would specifiy a time posterior to speech time, while the Past tense would specify the area of all the past time (relative to speech time) as a reference frame. The two tensing operations would contradict each other, the sentence would be uninterpretable. The version (21b), on the other hand, according to Fabricius-Hansen allows an interpretation, if the Past tense can be anchored to a contextually supplied time, e.g., the event time specified in a preceding sentence. Let us take the following text as an example:

(22) (A) He awoke early in the morning and started up.
(B) The ship left in 20 minutes.

The event time specified in (A) supplies an anchor (or evaluation time) for the Past tense in (B), which, according to Fabricius-Hansen's analysis of this tense in narrative contexts, will supply a reference frame positioned "non-homogeneously posterior" to the event time of (A) (which also marks the time of speech or thought of the narrated person). And the next lower temporal expression, i.e. *in 20 minutes*, will narrow down this reference frame to a time span round "twenty minutes later". In this interpretation the sentence could be given a truth-conditional analysis. Underlying this analysis is the assumption of two contrary functions for the Past tense: anteriority relative to the speech time in isolated sentences and posteriority relative to a contextually given evaluation time in contextually related sentences.

Her approach, however, raises several questions: First of all, Fabricius-Hansen takes it for granted that deictic adverbials such as *tomorrow* can be anchored to a contextually supplied reference time in the past without indicating that this time constitutes a secondary speech time, the time of speech, thought or perception for the narrated person, i.e. a secondary deictic field. This information, while a central part of the meaning of substitutionary speech and perception, vanishes in Fabricius-Hansen's approach, which, as it stands, may be truth-conditionally correct.

Secondly, the greatest handicap of this approach is the fact that the formulae proposed as bases for the semantic (truth-functional) analysis would in many cases fail to distinguish between narrative report and substitutionary speech. For our interests at least, these are indispensable.

Fabricius-Hansen discusses the following sequence indexed in her manuscript as (55'):

(55') (A) Am 6. April wurde Anne früh wach.
(B) Sie hatte viel zu tun,
(C) denn morgen kamen ihre Eltern zu Besuch.[61]

She says that (C) can have only one interpretation, an interpretation in which the Past tense must constitute the highest temporal predicate, the alternative configuration with the time adverbial as highest temporal predicate leading to a contradictory statement. But this contradiction will only ensue if, as in our case, the Past tense is combined with a future time adverbial. In the text under discussion both (C) and (B) can be looked upon as substitutionary speech: *Sie hatte viel zu tun*. In Fabricius-Hansen's approach the only analysis possible would be to anchor the Past tense of this sentence to the event time supplied by the preceding sentence (*-wurde Anne früh wach)* and let it establish a reference time "non-homogeneously posterior" to this time, i.e., an interpretation in the context (t_k, T_k, T_{ko}), where t_k is the event time of the preceding sentence and T_k the reference time established by the Past tense in (B). But this is the interpretation the Past tense receives in contextually related (i.e. narrative) sentences . It would not contain the slightest indication that it can be or is to be interpreted as substitutionary speech. Fabricius-Hansen's formula for this sentence - "Prät. (6. April (sie viel zu tun haben))" - would be unable to reflect the mode of speech, thought, and perceptual representation as it is indicated in the text.

5.3 Our own notational system: an apparatus for the analysis and reconstruction of the text

As outlined in Schopf 1984 (Chapter 7), our prime interest is in developing an analytic apparatus which, applied to narrative texts, not only registers the tensing structure of the text, but also provides sufficient information for the reconstruction of the original text.

5.3.1 Substitutionary speech and perception involving two deictic fields

Any such attempt will have to take into account the fact that substitutionary speech (or indirect free speech) and direct and indirect interior monologue

61 Fabricius-Hansen (1984:97).

involve two orientation axes or deictic fields, the deictic field of the narrated character which is embedded into the deictic field of the narrator. Alternatively we could say that the temporal universe of the narrator, the primary temporal model, consisting of a primary event time (E), reference time (R), speech time (S) and adverbially denoted time (TA) embeds or is related to the temporal universe of the narrated character, the secondary temporal model, which contains a secondary event time (E'), reference time (R'), speech time (S') and time adverbial (TA').

5.3.2 The two deictic fields represented by embedding a secondary tense into a primary tense

The interpenetration of two deictic fields could, of course, in a first approximation, be indicated by embedding one tense into another tense. For sentence (B) in (55') above this could be done by means of the formula

Pret.(Pres.(ich viel zu tun haben)).

If the highest temporal predicate in this formula, (Pret.), is understood as "it was the case that a simultaneity relation was expressed", the sentence could be interpreted as indicating substitutionary speech, triggering the transposition of the secondary tense (i.e. Present into the Past), of the personal pronoun *ich* into *er* or *sie*, but preserving the deictic time adverbial (if present), yielding *Sie hatte viel zu tun.*

5.3.3 The tenses supplanted by temporal relations

But such formulae, if used as the basis for reconstructing an original text in practical language teaching, require of the student only that he know the transposition rules for the tenses and pronouns. A more comprehensive knowledge of the English tensing system is required if we substitute temporal relations for the tenses as indicated above and the time adverbial.

5.3.4 Reference time introduced by the existential operator

We suggested above that - the speech time (S) being given - the reference time be introduced by the existential operator ($\exists t$) and located on the time line by its relation to speech time (S), or reference time (R), and the time adverbial (TA), while the aspectual interpretation of the event was to be indicated by the inclusion of the event in the reference time ($R \supset E$) or of the reference time in the event ($R \subset E$).

5.3.5 Reichenbach's three parameters as constants

The difficulty with this approach lies in the fact that the existential operator ($\exists t$) introduces only indefinite times. Definite times could, of course, be expressly indicated by having the formulae accompanied by textually specified frame conditions giving S, R, E and S', R', E' as constants.

Substitutionary speech (e.g. *He would go to Berlin tomorrow*) could be formulated as follows: Let S, R, E be speech time, reference time and event time in the narrator's (or basic) temporal model, S', R', E' speech time, reference time and event time in the narrated character's (or secondary) temporal model with R=S' defining the relation of the models and E representing the speech act (or act of thinking or perception) introducing S', then the appropriate formula is:

$$(R < S) \wedge (R = S') \wedge \exists I_{TA} \left[(I_{TA} = S' + 1day) \wedge (S' < R') \wedge (I_{TA} \supset R' \supset E') \right].$$ [62]

The expression "I_{TA}" is to be read as "adverbially specified reference interval".

5.3.6 The representation of the progressive of achievement sentences

This formula raises two questions. The first is the representation of the aspectual interpretation of the situation ($R' \supset E'$), which is satisfactory for the present sentence, but would require a more detailed analysis for the progressive of achievements such as "(he) be winning the race", "(he) be reaching the top of the mountain", or "(she) be arriving at the station" etc. The representation of these progressives by the formula ($R \subset E$) is excluded by the fact that the lexically denoted event is a punctual phase which is not referred to by the progressive. The progressive here refers to the prephase leading up to the lexically denoted culmination point of the situation. The formula ($R \subset E$) is, therefore, misleading and requires modification, perhaps by saying that it is the prephase of E that includes (and overlaps) R:

62 This procedure was introduced into our discussions by Cornelia Hamann.

He was winning the race

$(R_{(in)def} < S) \wedge (R \subset \text{prephase } E_{(he\ win\ the\ race)})$.

5.3.7 Result of listener's calculation not to be represented in the formulae

The second difficulty presented by our formula above is that it includes the result the listener is expected to arrive at with respect to the event's location on the time line by calculating the temporal relations expressed in the sentence. The expression $(I_{TA} \supset R' \supset E')$ is part of this calculation. The information described by it is not expressed in the sentence, but is inferred by the listener from the signals contained in, or conveyed by, the sentence. We propose to restrict the formulae to the expression of the information contained in these signals. We further propose to seperate the interpretation of the time adverbial $(I_{TA} \supset R')$ from the aspectual interpretation of the event $(R' \supset E')$. Our formula for the sentence *He would go to Berlin tomorrow,* therefore, runs as follows:

$$(R < S)^1 \wedge (R = S')^2 \left[(I'_{(TA=S'+1day)} \supset R')^3 \wedge (S' < R')^4 \wedge (R' \supset E'_{(I\ go\ to\ Berlin)})^5\right].$$

5.3.8 The representation of the introductory clause: substitutionary speech with a tagged introductory clause

Our formula still lacks any information about the speech act, the act of thinking or perception introducing the secondary temporal universe, the secondary deictic field, i.e. we have no information about the nature of primary E. This event in some cases appears in the surface text as a tagged introductory verb of saying, thinking or perceiving, and must, therefore, be taken account of in our formula. Tagged *he thought*, if added to our sentence, changes our formula into

$$(R < S)^1 \wedge (R \supset E_{(he\ think)})^2 \wedge (R = S')^3 \left[(I'_{(TA:S'+1day)} \supset R')^4 \wedge (S' < R')^5 \wedge (R' \supset E'_{(I\ go\ to\ Berlin)})^6\right].$$

If the tagged introductory verb of thinking is not present in the surface text, however, as in the example being discussed, then this formula could be changed by the insertion of additional braces around parenthesis[2]:

$$(R < S')^1 \wedge \{(R \supset E_{(he\ think)})^2\} \ldots$$

5.3.9 Our formula paraphrased

The expressions in the numerically ordered parentheses can be paraphrased as follows:

$(R < S)^1$: Look for or assume an (in)definite reference time anterior to the speech time in the narrator's deictic field.

$(R \supset E_{(he\ think)})^2$: The act of thinking is contained in this primary reference time.

$(R = S')^3$: The primary reference time coincides with the secondary speech time and introduces the deictic field of the narrated character (=the embedding of the secondary deictic field into the primary deictic field).

$(I'_{(TA:S'+1day)} \supset R')^4$: The day after the day including the secondary speech time contains the secondary reference time (=time adverbial anchored to secondary speech time).

$(S' < R')^5$: Look for or assume a secondary (in)definite reference time later than the secondary speech time (=tense choice in the secondary deictic field).

$(R' \supset E'_{(I\ go\ to\ Berlin)})^6$: The event referred to in the secondary deictic field is contained in the secondary reference time (aspect choice in the secondary deictic field).

On the basis of this information the position of the secondary event can be calculated as follows: The event "I go to Berlin" is in its totality contained in a time interval which is a subinterval of the day following the day containing the secondary speech time, which coincides with the primary event (he think) which is anterior to the primary speech time, i.e. the speech time of the narrator.

5.3.10 Our formula as an instruction for the reconstruction of the text

Our formula can also be used as an instruction for the reconstruction of the original sentence.

We introduce the convention that the reconstruction starts with the parenthesis with the highest index, i.e. with $(R' \supset E'_{(I\ go\ to\ Berlin)})^{6}$, which says that the proposition constituting the secondary event (E') is to be rendered in the simple form: *I go to Berlin.*

The parenthesis with the next highest index $(S' < R')^{5}$ is an instruction to choose the tense for the secondary event, which is posterior to the secondary speech time. The *shall/will* auxiliary suggests itself here: *I shall go to Berlin.*

The next parenthesis $(I'_{(TA=S'+1day)} \supset R')^{4}$ informs us that the secondary reference time is contained in a time interval which is the day following the secondary speech time, i.e. is to be lexicalized by *tomorrow: I shall go to Berlin tomorrow.*

Parenthesis $(R = S')^{3}$ embeds the secondary deictic field into the deictic field of the narrator, i.e. triggers the transposition of the grammatical person (*I* to *he*) and the tense (Future tense into the Conditional and *shall* into *will*): *He would go to Berlin tomorrow.* Please note that the deictic time adverbial is as a rule preserved.

The next parenthesis $(R \supset E_{(he\ think)})^{2}$ introduces the act of thinking as a tagged introductory clause.

According to the information in the last parenthesis $(R < S)^{1}$, the Past tense has to be chosen: *He would go to Berlin tomorrow, he thought.* The additional braces around parenthesis^{2} change the sentence into *He would go to Berlin tomorrow.*

Our formula seems to supply all the information necessary to calculate, on the one hand, the secondary and primary event times and to reconstruct, on the other, the original sentence, which presupposes familiarity with the rules and regularities of tense and aspect choice in English, the nature and use of the adverbials, the event notions, and the transposition rules for the more complex modes of narration.

5.4 The various techniques of speech and thought representation in our formulae

So far, we have only looked at sentences rendered in substitutionary speech with a tagged introductory clause. How can we represent the other modes of speech and thought representation?

5.4.1 Substitutionary speech without an introductory clause

Substitutionary speech without a tagged introductory clause can be indicated by including parenthesis[2] in additional braces:

He would go to Berlin tomorrow

would then be rendered by

$$(R < S)^1 \wedge \{(R \supset E_{(he\ think)})^2\} \wedge (R = S')^3 [(I'_{(TA:S'+1day)} \supset R')^4 \wedge (S' < R')^5 \wedge (R' \supset E'_{(I\ go\ to\ Berlin)})^6]$$

5.4.2 Direct and indirect interior monologue

Direct and indirect interior monologue lack transposition of the tenses, but whereas direct interior monologue dispenses with the transposition of the pronouns, indirect interior monologue subjects the personal pronouns to transposition. We interpret this as two grades of suppression of the deictic field of the narrator and indicate this by placing parenthesis[3] (which signals the embedding of the secondary deictic field into the primary one) between braces and angled brackets, thus

He will go to Berlin tomorrow

as indirect interior monologue will be rendered by

$$(R < S)^1 \wedge \{(R \supset E_{(he\ think)})^2\} \wedge \{(R = S')^3\} \ldots$$

and

I shall go to Berlin tomorrow

as direct interior monologue will require the addition of angled brackets around parenthesis[3]:

$$(R < S)^1 \wedge \{(R \supset E_{(he\ think)})^2\} \wedge \{<(R = S')^3>\} \ldots$$

5.4.3 The representation of mixed techniques

Our procedure also gives us a means of representing "mixed" narrative techniques. Leech and Short point out that free indirect speech (or substitutionary speech), instead of making use of the deictic time adverbial (e.g. *tomorrow*), can retain the corresponding relative time adverbial (e.g. *the following day*). They illustrate this with the following three versions rendering the same state of affairs[63]:

(a) He would return there to see her again the following day.
(b) He would return there to see her again tomorrow.
(c) He would come back there to see her again tomorrow.

Version (a) and (b) differ in their choice of the time adverbial. The latter uses the deictic time adverbial, whose most essential feature is that it is anchored to the secondary speech time in the deictic field of the narrated character: $(I'_{(TA:S'+1day)} \supset R')^4$. That version (a) makes use of the relative time adverbial can be indicated in parenthesis^4 by anchoring it to the primary reference time: $(I'_{(TA:\,R\,+1day)} \supset R')^4$.

Version (b) and (c) differ in their lexical material. The choice of "return there" reflects the point of view of the narrator, while "come back there" suggests the point of view of the narrated character. This difference cannot be represented by our formulae. It can only be indicated in the wording of the two propositions.

5.5 The Past tense in narrative texts

Let us now return again to the Past tense. The Past tense in the narrative surface text can have several sources.

5.5.1 The Past tense representing the original Past of the narrator

In narrative report it renders the original Past of the narrator.

5.5.2 The Past tense representing the Present tense of the narrated character in substitutionary speech and perception

In substitutionary speech and substitutionary perception it can be said to represent the Present tense of the narrated character. The following sequence of sentences contains both Past tenses:

---

63 Leech/Short (1982:325).

(A) He opened the blind.
(B) Snow was falling.

It would be represented as follows:

$$(A):\ (R_1 < S)^1 \wedge (R_1 \supset E_{(he\ open\ the\ blind)})^2$$
$$(B):\ (R_2 < S)^1 \wedge \{(R_2 \supset E_{(he\ see)})^2\} \wedge (R_2 = S')^3\ [(S' = R')^4 \wedge (R' \subset E'_{(snow\ fall)})^5].$$

5.5.3 The Past tense representing the Past tense of the narrated character in indirect interior monologue

Finally, the surface Past tense in narrative prose can be the untransposed Past of the narrated character in indirect interior monologue as, for instance, in the following passage from James Joyce's *A Portrait of the Artist as a Young Man:*

> They stood on the steps of the tram,
> he on the upper, she on the lower.
> She came up to his step many times
> between their phrases and went down again
> and once or twice remained beside him
> forgetting to go down and then went down:
> Let be! Let be![64]

This passage is not narrative report of events by the narrator, it is the narrated character reliving in his memory a scene he experienced years ago. It is Stephen speaking. The only indication that this interpretation is correct are the traces of proximal deixis in the lexicalization of this passage *((she) came up (to his step many times))* and perhaps the fact that the passage slides off into free indirect thought or direct interior monologue: *Let be! Let be!*

The first sentence of the passage quoted above would have to be represented by the following formula:

$$(R < S)^1 \wedge \{(R \supset E_{(he\ remember)})^2\} \wedge \{(R = S')^3\}\ [(R' < S')^4 \wedge (R' \subset E'_{(we\ stand...)})^5].$$

Parenthesis[5] says that the event "we stand on the steps..." surrounds the secondary reference time, which would normally lead to the progressive in contem-

64 Joyce (repr. 1980:201).

porary English, but the simple form is also permissible as a stylistic variant: *we stand on the steps....* Parenthesis[4] locates the event in the past of the narrated character (i.e. in the secondary temporal universe): *we stood on the steps....* Parenthesis[3] defines the relation of the secondary temporal universe to the universe of the narrator, i.e. the technique of presenting the narrated character's consciousness: Indirect interior monologue transposes only the pronominal reference: *they stood on the steps....*

5.5.4 How to represent the progression of reference time

There remains the question of how to represent the progression of reference times. We can only touch upon this subject in this paper. The question requires extensive and detailed investigation.

It seems to be generally accepted that stories are temporally organized via temporally related reference times, which need not necessarily follow each other in immediate succession. We interpret stories, biographies etc. against the background of our knowledge of the reality surrounding us. If we say, for instance, that "somebody was born in Prague and went to school in Vienna", we are not trying to report two events in immediate succession: Stories, biographies etc. step through time. The interpretation of a narrative text presupposes a familiarity with extra-linguistic reality and the human experience of the temporal organization of the universe. Our interpretation of stories permits and assumes a good deal of vagueness in the temporal relation of reference times.

Another important aspect of the temporal organization of narrative texts is the fact that the more complex narrative techniques such as substitutionary speech and perception etc. represent two temporal universes and a sequence of events (and reference times) in each of them. Our procedure accounts for this fact by enabling us to represent a primary series of reference times ($R_1 - R_2 - R_3$) to be related to a secondary series of reference times ($R_1' - R_2' - R_3'$ etc.) via an embedding relation ($R = S'$). We indicate these two levels of reality by including the tensing operations of the narrated character in square brackets.

5.5.5 How to account for the progression of reference time

Finally, there is the question of how to account for the progression of reference times in narrative texts. As pointed out above (p.196), we do not believe that any kind of automatic progression of the reference time triggered by event sentences represents a realistic picture of the way narrative texts are processed by the listener or reader. Instead, we propose that the processing of narrative texts should be looked upon as a calculation procedure in which every new sentence or clause is evaluated against the background of (all the) information supplied by the preceding text (or co-text) and semantic and pragmatic regularities supplied by the language system in question and our knowledge of extra-linguistic reality. One of the basic rules seems to be the instruction to identify the reference time of a newly introduced sentence with the reference time via which the preceding sentence was tensed, unless this is prevented by semantic or pragmatic temporal incompatibility of the two events reported in the two sentences. Further research is necessary to find answers to these questions.

6.0 Peculiarities of tense and aspect in English

In the present paper, we have not been able to present more than a fragmentary picture of the English Tensing system. To describe exhaustively even one subcomponent of this, such as tense and aspect, would require much more space than is available. We may, however, mention one or two peculiarities of this system that pose questions for future research.

6.1 The functional contrast between simple and progressive verb forms not exhaustively described

The functional contrast between the simple and the progressive verb form in English can be understood as an aspectual contrast, at least in the Present and Past tenses. The simple Past can in many cases be understood as predicating the event in question as a whole or a blob, i.e. as contained in the reference time. In our formulae this is indicated by having the reference time include the event ($R \supset E$), while the progressive Past can be interpreted as the event including or surrounding the reference time ($R \subset E$). This is not an exhaustive functional analysis of the aspect opposition in English. Besides the basic meaning difference between these two forms as just mentioned both the progressive and the simple form probably have to be accorded several secondary functions.[65]

65 Cf. Comrie (1976:24ff.); Schopf (1984:261ff.).

6.2 Present Perfect and Past in English

Probably the most outstanding peculiarity of the tense system in English is the contrast between the Present Perfect and the Past tense.

The Past tense was interpreted above as a search instruction, i.e. as an instruction to look for an (in)definite past reference time either identical with the event time or in the case of states and unquantified processes included in the event time, or for a definite or indefinite event time in a specified or presupposed (adverbially defined) reference frame. This description seems to provide an explanation of the collocational regularities for the combination of the Past tense with time adverbials. It further explains why sequences of sentences in the Past tense can denote sequentially ordered events and in some cases also sequentially unordered events. The precise conditions, however, under which these meanings can be conveyed are not yet fully known.

The Present Perfect differs from the Past tense in that it can be understood as an instruction to look, proceeding from speech time, for an indefinite event time within a reference frame reaching up to and including speech time. This description of the Present Perfect explains why events reported in this tense are, as a rule, not sequentially ordered: *I have read, played tennis, gone for long walks but nothing has helped*. The events "read", "play tennis", and "go for walks" are not, or need not be, sequentially ordered.[66]

6.3 The Pluperfect and time adverbials

Another peculiarity of the English tense system is the fact that the Pluperfect can be combined with a time adverbial specifying the reference time in relation to which the predicated event is anterior and a time adverbial specifying the event time itself: *When you rang, the last customer had left several minutes before*. But the Pluperfect can also refer to an indefinite event time: *When I looked into his room he had fallen asleep*. The time of the event "fall asleep" need not be related to a definite event time.

6.4 The uses of the uncoloured Future in English

A third peculiarity of the English tense system is its structural dissymmetry. While there are three "past" tenses: the Pluperfect, the Past , and the Present

---

66 Cf. Matthews (this volume).

Perfect, there are, assuming that uncoloured future reference can be brought about by *shall/will* plus an aspectually unmarked infinitive, only two tenses for the future: the ordinary Future and the Future Perfect. If we assume that the Pluperfect corresponds to the Future Perfect, then the ordinary Future corresponds to the Past and Present Perfect. We shall consequently have to ask whether the ordinary Future comprises for its sphere the functions corresponding to the functions of both the Past and Present Perfect tense, i.e. whether we have future reference anchored to a reference time coinciding with the speech time.

There is no doubt that the ordinary Future can relate an event to a definitely specified time in the future: it can be combined with definite time adverbials that are identical with the event time: *The train will depart at 12 o'clock*. Moreover, it can certainly refer to a definite time within a definite future reference frame: *Father will come back from Paris next week*. Here the speaker may have a definite time in his or her mind. But the Future tense may also refer to an indefinite event time within a definite reference frame: *He will certainly be able to get out of bed next week*. Here the speaker need not have a definite time in mind and his utterance will have to count as true if the event happens at any time 'next week'.

The uses of the ordinary Future tense so far discussed correspond to those of the Past tense. The question that remains to be answered is whether it has uses corresponding to those of the Present Perfect tense. I believe it does. First of all there is an ordinary Future corresponding to the continuative perfect: *I have waited for an hour - We shall wait here for an hour*, and there is it seems an ordinary Future that refers to an indefinite time in the future: *He will turn out to be a traitor*.This sentence has to be taken as true if there is <u>some</u> time in the future at which the event predicated occurs. Can we say that this sentence refers to an indefinite future event time from a reference time coinciding with the speech time: R,S——E?

6.5 Tense forms containing two predicates

Attention should also be paid to tense forms that contain two predicates that can be seperately related to time adverbials. *Yesterday Mother was coming tomorrow* is an interpretable English sentence. The verb form or tense form contains a modal component which can be described as "intention" and a predicate naming an event, "come". The first time adverbial specifies a reference frame for the intention, the second a time for the intended event. There may be several other tense uses of this kind. We must leave these too to further studies.

Bibliography

Abraham, W. (1980) The Synchronic and Diachronic Semantics of German Temporal 'noch' and 'schon', with Aspects of English 'still', 'yet' and 'already', Studies in Language 4, 3-24.

Allen, R.L. (1966) The Verb System of Present-Day American English. The Hague: Mouton.

Åquist, L. (1976) Formal Semantics for Verb Tenses as Analysed by Reichenbach, in T. A. van Dijk (ed.) Pragmatics of Language and Literature. Amsterdam: North Holland Publishing Co., 229-236.

Åquist, L., F. Guenthner and C. Rohrer (1977) 'Soon' and 'recently', in C. Rohrer (ed.) (1977), 67-81.

Bach, E. (1981) On Time, Tense and Aspect. An Essay in English Metaphysics, in P. Cole (ed.) Radical Pragmatics. New York: Academic Press, 63ff.

Bäuerle, R. (1979) Temporale Deixis, temporale Frage. Zum propositionalen Gehalt deklarativer und interrogativer Sätze. Tübingen: Narr.

Bäuerle, R., V. Egli and A. von Stechow (eds.) (1979) Semantics from different points of view. Berlin: Springer.

Bauer, G. (1970) The English 'Perfect' reconsidered, Journal of Linguistics 6, 189-198.

Bennett, M. and B. Partee (1978) Toward the Logic of Tense and Aspect in English (manuscript reproduced by the Indiana University Linguistics Club).

Brinkmann, F. (1884/85) Syntax des Französischen und Englischen in vergleichender Darstellung (2 vols.). Braunschweig: Vieweg.

Bronzwaer, W.J.M. (1970) Tense in the Novel. An investigation of some potentialities of linguistic criticism. Groningen: Wolters-Norhoff.

Bull, W.E. (1968) Time, Tense and the Verb. A Study in theoretical and applied linguistics, with particular attention to Spanish. Berkeley and Los Angeles: University of California Press.

Chafe, W. (ed.) (1980) The Pear Stories. Cognitive, cultural, and linguistic aspects of narrative production. Norwood, N.J.: Ablex.

Close, R.A. (1975) A Reference Grammar for Students of English. London: Longman.

Comrie, B. (1976) Aspect. An Introduction to the Study of Verbal Aspect and Related Problems. Cambridge: Cambridge University Press.

Comrie, B. (1985) Tense. Cambridge: Cambridge University Press.

Comrie, B. (1986) Tense and time reference: from meaning to interpretation in the chronological structure of a text, Journal of Literary Semantics 15, 12-22.

Couper-Kuhlen, E. (1987) On the temporal interpretation of postposed 'when'-clauses in narrative discourse, in R. Matthews and J. Schmole-Rostosky (eds.) Papers on Language and Mediaeval Studies presented to Alfred Schopf - on the occasion of his 65th birthday. Frankfurt/M.: Peter Lang.

Couper-Kuhlen, E. (in preparation) Temporal relations and foregrounding in narrative discourse, in A. Schopf (ed.) (in preparation) Essays on Tensing in English, Vol. II: Time, Text and Modality. Tübingen: Niemeyer.

Cresswell, M. J. (1973) Logics and Languages. London: Methuen.

Crystal, D. (1966) Specification and English Tenses, Journal of Linguistics 2, 1-34.

Dahl, Ö. (1984) Temporal distance: remoteness distinctions in tense-aspect systems, in B. Butterworth, B. Comrie, Ö. Dahl (eds.) Explanations for Language Universals. Berlin: Mouton, 105-122.

Dietrich, G. (1955) Erweiterte Form, Präteritum und Perfektum im Englischen. (Eine Aspekt- und Tempusstudie.) München: Hueber.

Dinsmore, J. (1982) The Semantic Nature of Reichenbach's Tense System, Glossa 1982, 216-238.

Dowty, D. R. (1979) Word Meaning and Montague Grammar. The semantics of verbs and times in generative semantics and in Montague's PTQ. Dordrecht: Reidel.

Dowty, D. R. (1982) Tenses, Time Adverbs, and Compositional Semantic Theory, Linguistics and Philosophy 1982, 23-59.

Enkvist, N. E. (1981) Experiential iconicism in text strategy. Text 1, 97-111.

Fabricius-Hansen, C. (1984) Tempus fugit. Über die Interpretation temporaler Strukturen im Deutschen (manuscript). Oslo: Germanistics institutt.

Fehr, B. (1938) Substitutionary narration and description. A chapter in stylistics, English Studies 20, 97-107. Also in B. Fehr (1944), 264-279.

Fehr, B. (1944) Von England's geistigen Beständen. Frauenfeld: Huber.

Fillmore, C.F. (1975) Santa Cruz Lectures on Deixis (delivered 1971). Indiana University Linguistics Club.

Galton, A. (1984) The Logic of Aspect. An axiomatic approach. Oxford: Clarendon Press.

Garey, H. B. (1957) Verbal Aspect in French, Language 33, 91-110.

Guenthner, F. (1979) Time Schemes, Tense Logic and the Analysis of English Tenses, in F. Guenthner and S.J. Schmidt (eds.) Formal Semantics and Pragmatics for Natural Languages. Dordrecht: Reidel, 201-222.

Halliday, M. A. K. (1976) The English Verbal Group, in G. Kress (ed.) System and Function in Language: Selected Papers by M. A. K. Halliday. London: Oxford University Press, 136-158.

Harkness, J. (1985) On the Semantic Properties of English Time Adverbials. (Diss.) Freiburg.

Harris, J. (1984) Syntactic Variation and Dialect Divergence, Journal of Linguistics 20, 303-327.

Harris, M. (1986) Aspects of subordination in English and other languages. Bulletin of the John Rylands University Library of Manchester, Vol. 69/1.

Harris, M. (forthcoming) Markedness and clause structure, in O. M. Tomic (ed.) Markedness in Synchrony and Diachrony. Berlin: Mouton de Gruyter.

Heinämäki, O. (1978) Semantics of English Temporal Connectives. Indiana University Linguistics Club.

Hinrichs, E. (1981) Temporale Anaphora im Englischen. (Zulassungsarbeit) Tübingen.

Hirtle, W. H. (1977) Already, Still and Yet, Archivum Linguisticum, Vol. VIII N. S. No. 1, 28-45.

Huddleston, R. (1984) Introduction to the Grammar of English. Cambridge: Cambridge University Press.

Isard, S. D. (1974) What would you have done if...?, Theoretical Linguistics (1974), 233-255.

Jacobson, S. (1964) Adverbial Positions in English. (Diss. Uppsala) Uppsala: Ab Studentbok, SFS:s bokförlag.

Jespersen, O. (1924) The Philosophy of Grammar. London: Allan and Unwin.

Jespersen, O. (1933) Essentials of English Grammar. London: Allan and Unwin.

Jessen, M. E. (1974) A Semantic Study of Spatial and Temporal Expressions in English. (Diss. Edinburgh).

Joos, M. (1964) The English Verb: Form and Meanings. Madison: University of Wisconsin Press.

Joyce, J. (1977) A Portrait of the Artist as a Young Man. London: Granada Panther Book (repr. 1980).

Kamp, H. (1971) Formal Properties of "Now", Theoria 1971, 227-273.

Kamp, H. (1975) Two Theories about Adjectives, in E. L. Keenan (ed.) Formal Semantics of Natural Language. Cambridge: Cambridge University Press, 123-255.

Kamp, H. (1979) Events, instants and temporal reference, in R. Bäuerle, U. Egli and A. von Stechow (eds.) (1979), 376-417.

Kamp, H. (1981) A Theory of Truth and Semantic Representation, in J. Groenendijk, T. Janssen, M. Stokhof (eds.) Formal Methods in the Study of Language. Part 1. Mathematisch Centrum Amsterdam, 277-322.

Kamp, H. and C. Rohrer (1983) Tense in Texts, in R. Bäuerle, C. Schwarze, A. von Stechow (eds.) Meaning, Use, and Interpretation of Language. Berlin: de Gruyter, 250-269.

Kiefer, F. (to appear) Temporal structure as a linking device, in F. Kiefer (ed.) Linking in Text. Dordrecht: Reidel.

Kilby, D. (1984) Descriptive Syntax and the English Verb. London: Croom Helm.

Koschmieder, E. (1945) Zur Bestimmung der Funktionen grammatischer Kategorien, in: Abhandlungen der Bayerischen Akademie der Wissenschaften, Neue Folge, Bd. 25, 22-29. München.

Labov, W. (1972) The transformation of experience in narrative syntax, in: Language in the Inner City. Studies in the Black English vernacular. Philadelphia: University of Pennsylvania Press, 354-396.

Labov, W. and J. Waletzky (1967) Narrative analysis: Oral versions of personal experience, in J. Helm (ed.) Essays on the Verbal and Visual Arts. Seattle: University of Washington Press, 12-44.

Leech, G. N. (1969) Towards a Semantic Description of English. London: Longman.

Leech, G. N. (1971) Meaning and the English Verb. London: Longman.

Leech, G. N. and J. Svartvik (1975) A Communicative Grammar of English. London: Longman.

Leech, G. N. and M. H. Short (1981) Style in Fiction: A Linguistic Introduction to English Fictional Prose. London: Longman.

Levinson, S. C. (1983) Pragmatics. Cambridge: Cambridge University Press.

Lewis, D. (1979) Scorekeeping in a Language Game, in R. Bäuerle, U. Egli, A. von Stechow (eds.) (1979).

Lyons, J. (1977) Semantics. Vol. II. Cambridge: Cambridge University Press.

Matthiessen, C. and S. A. Thompson (forthcoming) The structure of discourse and "subordination", in J. Haiman and S. A. Thompson (eds.) Clause-combining in Grammar and Discourse. Amsterdam: Benjamins.

McCawley, J. D. (1971) Tense and Time Reference in English, in C. J. Fillmore and D. T. Langendoen (eds.) Studies in Linguistic Semantics. New York: Holt Rinehart & Winston, 97-113.

McCawley, J. D. (1981) Everything that Linguists have Always Wanted to Know about Logic (but were ashamed to ask). Oxford: Blackwell.

McCoard, R. W. (1978) The English Perfect: Tense-Choice and Pragmatic Inferences. Amsterdam: North Holland Publishing Co.

Montague, R. (1974) The Proper Treatment of Quantification in Ordinary English, in R. Thomason (ed.) Formal Philosophy. Selected Papers of R. Montague. New Haven: Yale University Press.

Nerbonne, J. (1982) Reference time and time in narration. Paper presented at the Sloan Foundation Conference on Tense and Aspect in Discourse, May 1982, Ohio State University.

Nerbonne, J. (1984) German Temporal Semantics: Three-dimensional tense logic and a GPSG fragment. (Diss. Ohio State University). Working Papers in Linguistics 30, Department of Linguistics, Ohio State University (also available as Dissertation ts. 1983).

Onions, C. T. (1971) Modern English Syntax. (New edition by B. D. H. Miller.) London: Routledge & Kegan Paul.

Palmer, F. R. (1974) The English Verb. London: Longman.

Palmer, H. E. and F. G. Blandford (1939) A Grammar of Spoken English (2nd edition). Cambridge: Heffer.

Palmer, H. E. and F. G. Blandford (revised and rewritten by R. Kingdon) (1969) A Grammar of Spoken English (3rd edition). Cambridge: Heffer.

Partee, B. Hall (1973) Some structural analogies between tenses and pronouns in English, Journal of Philosophy 70, 601-609.

Partee, B. (1984) Nominal and Temporal Anaphora, Linguistics and Philosophy 7, 243ff.

Prior, A. (1967) Past, Present and Future. Oxford: Oxford University Press.

Quirk, R., S. Greenbaum, G. Leech and J. Svartvik (1972) A Grammar of Contemporary English. London: Longman.

Quirk, R., S. Greenbaum, G. Leech and J. Svartvik (1985) A Comprehensive Grammar of the English Language. London: Longman.

Rauh, G. (1978) Linguistische Beschreibung deiktischer Komplexität in narrativen Texten. Tübingen: Narr.

Rauh, G. (1984a) Aspekte der Deixis I: Deiktische Determination. L. A. U. T., Series A, No. 122.

Rauh, G. (1984b) Aspekte der Deixis II: Deiktische Dimensionen und die Verwendung deiktischer Ausdrücke. L. A. U. T., Series A, No. 123.

Reichenbach, H. (1947) Elements of Symbolic Logic. New York: The Macmillan Co. (repr. 1980).

Reinhart, T. (1984) Principles of gestalt perception in the temporal organization of narrative texts, Linguistics 22, 779-809.

Rescher, N. and A. Urquart (1971) Temporal Logic. Wien and New York: Springer.

Ritchie, G. D. (1979) Temporal clauses in English, Theoretical Linguistics 1979, 87-115.

Rohrer, C. (ed.) (1977) On the Logical Analysis of Tense and Aspect. Tübingen: Narr.

Saurer, W. (1984) A Formal Semantics of Tense, Aspect, and Aktionsarten. Bloomington: Indiana University Linguistics Club.

Schmole, J. P. (1977/8) Strukturen der Aspektwahl im neuenglischen Perfekt. Teil I in: Kritikon Litterarum 6, 240-254. Teil II in: Kritikon Litterarum 7, 89-101.

Schopf, A. (1976) Lexikalische Klassen als Grundlage für die Beschreibung des englischen Verbalsystems, Anglia 94, 1-43.

Schopf, A. (1984) Das Verzeitungssystem des Englischen und seine Textfunktion. Tübingen: Niemeyer.

Smith, C. S. (1978) The Syntax and Interpretation of Temporal Expressions in English, Linguistics and Philosophy 2, 43-99.

Smith, C. S. (1980) Temporal Structures in Discourse, in C. Rohrer (ed.) Time, Tense, and Quantifiers. Proceedings of the Stuttgart Conference on the Logic of Tense and Quantification. Tübingen: Niemeyer, 355-374.

Smith, C.S. (1981) Semantic and Syntactic Constraints on Temporal Interpretation, in P. Tedeschi and A. Zaenan (eds.) Tense and Aspect. (Syntax and Semantics, vol. 14). New York: Academic Press, 213-237.

Sørensen, H. (1964) On the Semantic Unity of the Perfect Tense, in: English Studies Presented to R.W. Zandvoort on his 70th Birthday. Amsterdam: Swets & Zeitlinger, 74-83.

Steinberg, G. (1971) Erlebte Rede. Ihre Eigenart und ihre Formen in neuerer deutscher, französischer und englischer Erzählliteratur. Göppingen: Kümmerle.

Sweet, H. (1900) A New English Grammar: Logical and Historical. (2 vols.) Oxford: Oxford University Press.

Tichý, P. (1980) The Logic of Temporal Discourse, Linguistics and Philosophy 1980, 343-369.

Traugott, E. C. and J. Waterhouse (1969) 'Already' and 'yet': a suppletive set of aspect-markers, Journal of Linguistics 5, 287-304.

Twaddell, W. (1963) The English Verb Auxiliaries. Providence: Brown University Press.

Vendler, Z. (1957) Verbs and times, Philosophical Review 66, 143-160. Also in A. Schopf (ed.) (1974) Der englische Aspekt. Darmstadt: Wissenschaftliche Buchgesellschaft, 217-234.

Wiener, N. (1914) A Contribution to the Theory of Relative Position, in: Proceedings of the Cambridge Philosophical Society, Vol. 17.

Wolfson, N. (1979) The conversational historical present alternation, Language 55, 168-182.

Wright, R. (1972) Native Son. Harmondsworth: Penguin Books.

Zandvoort, R. W. and J. A. van Ek (1975) A Handbook of English Grammar. (7th edition). London: Longman.

Zydatiß, W. (1976) Tempus und Aspekt im Englischunterricht. Kronberg/Ts.: Scriptor.

Bei Fragen zur Produktsicherheit wenden Sie sich bitte an:
If you have any questions regarding product safety,
please contact:

Walter de Gruyter GmbH
Genthiner Straße 13
10785 Berlin
productsafety@degruyterbrill.com